J. B. Wheeler

Elements of field fortification for the use of the cadets of the United States military academy at West Point, N.Y. by J. B. Wheeler

J. B. Wheeler

Elements of field fortification for the use of the cadets of the United States military academy at West Point, N.Y. by J. B. Wheeler

ISBN/EAN: 9783743347748

Manufactured in Europe, USA, Canada, Australia, Japa

Cover: Foto ©berggeist007 / pixelio.de

Manufactured and distributed by brebook publishing software (www.brebook.com)

J. B. Wheeler

Elements of field fortification for the use of the cadets of the United States military academy at West Point, N.Y. by J. B. Wheeler

THE ELEMENTS OF FIELD FORTIFICATIONS

FOR THE USE OF THE CADETS OF THE UNITED STATES MILITARY ACADEMY, AT WEST POINT, N. Y.

BY

J. B. WHEELER,

PROFESSOR OF CIVIL AND MILITARY ENGINEERING IN THE U. S. MILITARY ACADEMY AT WEST POINT, N. Y., AND BREVET COLONEL, UNITED STATES ARMY.

NEW YORK:
D. VAN NOSTRAND COMPANY,
23 MURRAY ST. AND 27 WARREN ST.
1893.

"There are five things from which the soldier must never be separated: his gun, his ammunition, his knapsack, his rations for four days and an intrenching tool."

NAPOLEON.

PREFACE.

This text-book is prepared for the use of the cadets of the United States Military Academy while pursuing their course of studies in the subject of "military engineering."

The endeavor has been made to state concisely and plainly the principles upon which the "art of fortification" is based, and to give all information likely to be of practical use to a young officer while serving in the field. All unnecessary details have been avoided, leaving explanations and illustrations of that kind to be introduced into the class-room.

The elementary form of the work and the method of treatment of the subject are based upon the assumption that the readers of the book are beginners and know nothing of the principles of fortification.

West Point, N. Y.. *February*, 1880.

CONTENTS.

CHAPTER I.

General Principles and Definitions.

CHAPTER II.

Elements of the Profile of a Fortification.

CHAPTER III.

Dimensions and Inclinations given to the Lines and Slopes of a Parapet.

CHAPTER IV.

The Trace of a Field Fortification.

CHAPTER V.

Field Works.

CHAPTER VI.

Lines.

Continued Lines.

Lines with Intervals.

CHAPTER VII.

The size of a Field Work, the number of its Garrison, and the Selection of its Trace.

CHAPTER VIII.

Construction of Field Works.

CHAPTER IX.

Revetments.

CHAPTER X.

Defilade.

CHAPTER XI.

The Interior Arrangements made in a Field Work.

Arrangement of the Parapet.

Arrangements for Sheltering the Troops, etc. from the effects of the Enemy's Fire.

Arrangements affording Communications, etc.

Arrangements for the Comfort and Health of the Garrison.

Other Interior Arrangements used.

CHAPTER XII.

Arrangements made Exterior to the Parapet.

ARTICLE. PAGE

Ditch Defences.

Obstacles.

CHAPTER XIII.

Application of Field Fortifications to Sites upon Irregular Ground.

Bridge Heads.

CHAPTER XIV.

Hasty Intrenchments.

CHAPTER XV.

Attack and Defence of Field Fortifications.

CHAPTER XVI.

Siege Works.

CHAPTER XVII.

Military Bridges.

CHAPTER XVIII.

Railroads.

INTRODUCTION.

INTRENCHMENTS are co-eval with man. In all ages they have played important parts in the contests carried on by force, when men have striven for the mastery.

They have been recognized by the great masters of the art of war as important adjuncts of their military operations, and have been freely used by them in their campaigns.

Napoleon said, "those who proscribe the help which the engineer's art may afford in the field, deprive themselves voluntarily of an auxiliary force and expedient, never hurtful, always useful, and often indispensable."

Frederick the Great, in his Mémoires, says,—"Officers require different kinds of knowledge, but one of the principal is that of fortification."

General Sherman says in his memoirs, "earth-forts, especially field-works, will hereafter play an important part in wars, because they enable a minor force to hold a superior one in check for a *time,* and time is a most valuable element in all wars."

The general of modern times who first prominently recognized the great merit of field works, and who saw in them "an auxiliary force," was the emperor, Charles V. of Germany.

To each regiment of infantry he attached a body of pioneers, numbering four hundred, under the command of a special officer, and provided them with intrenching tools. These men were used to construct fortifications, and perform the duties of engineer soldiers.

The use of this kind of force was fully appreciated by him, and on more than one occasion proved of great service to him.

It is mentioned, that in 1547, the emperor found himself in the presence of the enemy, who were twice his strength in numbers, but who made the mistake of not attacking at once. Charles V. immediately began to intrench his army, and, by morning of the following day, was capable of repulsing the enemy. He continued to strengthen his position for twelve days. On the thirteenth day, he received re-inforcements and moved against the enemy.

It is stated, that Peter the Great, at the battle of Pultowa, which terminated in the total defeat of the Swedish army, owed his success to intrenchments which he had had the foresight to construct in front of his line of battle, upon the night before the action.

The simple redoubt in the pass near Montenotte, defended by Colonel Rampon, in 1796, enabled Bonaparte to execute the details of a campaign unparalleled in the records of military history. Relative to this re doubt and its defence, Thiers says, "It saved the general's plans, and perhaps, the future of the campaign."

Intrenchments played an important part in the late war in the United States.

In the beginning of this war, (1861), the sentiment or general opinion of the American volunteers was adverse to intrenchments. The manual labor requisite for the construction of intrenchments was regarded by them as degrading; their idea was "a fair stand-up fight," in the open field. They applied the term "dirt-diggers" to the advocates of intrenchments, and considered "masked batteries" as devices unbecoming a civilized people.

Experience in war, and great necessity soon drove such erroneous views from their minds, and before the war ended, these same soldiers threw up intrenchments whenever the army halted, of their own free will and accord.

A writer stated, on this subject, that "they waited neither for orders, nor deployment of skirmishers, nor formation of lines. * * * * The rule was that the troops should proceed with this work without waiting for orders."

Their good sense soon caused them to appreciate the value of both natural and artificial aids in the defence, and to learn that the fate of a battle frequently depended upon some slight protection given to a line, a protection so slight in many cases as to be hardly worthy of the name, such as a few rails heaped together, a shallow trench, etc. The result was that the American soldier became an adept in intrenching himself on the field of battle.

The Confederates made great use of intrenchments, and by their aid inflicted great losses upon our troops.

General Grant crossed the Rapidan in May, 1864, and attacked the enemy in the Wilderness. His most desperate assaults were not successful in driving the confederates from their intrenchments. He was successful only by moving around the position.

The losses suffered by the Union army, under General Grant from the time he crossed the Rapidan until he reached the James river,—about one month,—were equal in numbers to the entire force opposed to him in the beginning, when he crossed the Rapidan; and these losses were largely due to the fact that the confederates occupied "intrenched" positions.

General Sherman bears testimony in many places to the influence of intrenchments. In speaking of the assaults made on the enemy's lines, at Kenesaw mountain, June 24, 1864, he says. "The two assaults were made at the time and in the manner prescribed, and both failed, costing us many valuable lives, * * * * our aggregate loss being near three thousand men, while we inflicted little loss upon the enemy who lay behind his well formed breast-works."

The experience acquired in this war in the United

States proved the necessity and value of intrenchments, even when extemporized and constructed in the presence of an enemy. It also proved that a very slight intrenchment, when manned, and with the approaches to it swept by artillery, was practically impregnable.

The cost of carrying a line of intrenchment is incidentally mentioned in one of the reports of Major General H. G. Wright, the present Chief of Engineers, United States Army, who commanded the Sixth Corps of the Army of the Potomac, in the attack upon Petersburgh, in April, 1865. He said that it "cost us, in killed and wounded, a number equal perhaps to that of the entire force of the enemy actually opposed to us. It was an attack of nearly two divisions against a picket line covered by a simple trench and parapet; but had it been held by two ranks of good troops it is doubtful if it could have been carried even by an entire corps." His conclusion was that "a well intrenched line, defended by two ranks of infantry cannot be carried by a direct attack," unless there are favorable circumstances, which it is not necessary to mention here.

The recent war between Russia and Turkey confirms the experience gained by us in our late war, and places the intrenching tool almost on a level with the musket.

Lieut. Greene in his report says, that General Skobeleff made his men carry the spades and shovels on their persons. He remarks, relative to this intrenching tool, "they were heavy (weighing over five pounds), they were uncomfortable, they were in every way inconvenient, but each man had learned by hard experience to feel that his individual life depended upon his musket and his spade—and he took good care to lose neither the one nor the other."

The student, seeing the advantages accruing from intrenchments hastily constructed, can easily draw his conclusions as to the value of lines of fortifications when there is time to build them properly.

The famous lines of Torres Vedras, in Portugal constructed by Lord Wellington; the line of works built around the city of Washington during the late war; the line of works defending Nashville, Tennessee; and others are interesting subjects of study, and show how a defeated, or an inferior army by their use is able to hold its ground, for more or less time, against a superior or victorious enemy.

Field fortifications have attained such great importance, owing to the long range, the precision, and the rapid fire of modern arms, that no soldier can afford to be without a knowledge of the great principles underlying this branch of military engineering. The remark of Frederick the Great, true in his day, is still true at the present time, viz: "Officers require different kinds of knowledge, but one of the principal is that of fortification."

FIELD FORTIFICATIONS.

CHAPTER I.

GENERAL PRINCIPLES AND DEFINITIONS.

1. Positions.—The term, **position,** is applied by military writers to any piece of ground which is, or may be, occupied by a body of troops for the purpose either of making or repelling an attack.

When the purpose is that of repelling an attack, the position is called a **defensive** one. The motives which lead a body of troops to occupy a defensive position may spring from several causes. The most prominent of these causes are smallness of numbers, inferiority in arms, lack of experience in war, or similar differences, which, at the time, render the enemy superior in strength.

A defensive position is occupied principally for the purpose of neutralizing this temporary superiority of an enemy, and to give to the troops holding the position, the chances of making a successful resistance to any attack made to dislodge them.

2. Strong positions.—Positions are called **strong**, when they give to their defenders **all** the chances of making a successful resistance against any sudden assault which may be made by an enemy.

When the natural features of a position are such, that the defenders can shelter themselves from the enemy's fire, and at the same time be free to pour a fire upon the enemy's columns as they approach, it is evident that the chances are decidedly in favor of a successful resistance by the defence, and the position is more or less a strong one. Positions are not equally strong, in consequence of the natural features of the ground in one case being less favorably disposed for shelter, than in another. Their strength may however be increased by artificial means, and so much so, in many cases, as to render them technically "strong" or impregnable.

3. Fortifications.—The artificial means used to give additional strength to a position are termed **fortifications**, from the verb, "to fortify" [*fortis*, strong, and *facere* to make] derived from the Latin.

The skilful combination of these means, and their construction, constitute a branch of Military Engineering known as the "**Art of fortification.**"

The object of the art of fortification is to locate and construct, upon positions selected for defence, such additional aids or helps, that the defenders can by their use remove partially, if not entirely, the disparity existing between them and their assailants.

The art of fortification may therefore be defined to be that branch of military engineering which has for its object **the strengthening of positions selected for defence.**

4. Classes of fortifications.—Fortifications are usually divided into two general classes, viz.:

Permanent and Temporary.

That class of fortifications which is usually constructed in times of peace for the purpose of strengthning positions which may be of military importance in case of a future war, is termed **permanent.** Fortifications of this class are built usually with great care, and of durable materials. They are expected to last for a considerable time, and are therefore comparatively permanent in their nature.

That class of fortifications which is usually built after war has been declared, and to strengthen positions which have suddenly acquired a military importance, is known as **temporary**. Fortifications of this class may in some cases become permanent, but ordinarily they are built in a comparatively short time, of materials which are near at hand and can be obtained quickly, and frequently by the labor of the troops. The positions strengthened by them are frequently of transitory importance and soon abandoned, and the materials used are oftentimes perishable in their nature. The fortifications are in use, in many cases, but for a short time; are wanting in durability; and are therefore temporary in their nature.

5. Field fortifications.—The temporary fortifications constructed by the labor of the troops constitute the greater part of the fortifications of this class. The troops are said to be "in the field" at this time, and the fortifications constructed by them are usually known as **field fortifications.**

The subject of this book is a general discussion of the principles used, and the kind of works employed, in fortifying a position in the field, by the labor of troops.

6. Kinds of field fortifications.—Fortifications constructed in the field are of two general kinds, according to the time employed in their construction, viz: **Hasty** and **Ordinary.**

Hasty fortifications are those which must be built in the time intervening between the end of a march and the beginning of a battle. This time is a very few hours at most, and the fortifications must be finished usually in a single night, or they will be useless.

Ordinary fortifications are also constructed hurriedly, but there is time enough to practically finish them according to the plans upon which they are laid out.

Hasty fortifications are the kind built, in many cases, in the actual presence of an enemy; ordinary fortifications are those built before the enemy arrives upon the ground.

7. General principles of fortifications.—Fortifications, whatever be their class, are merely passive means of defence.

Certain general conditions must be satisfied by a defensive position, to enable an armed force occupying it to contend successfully against an assailant, superior either in numbers, discipline, or arms. These conditions must be favorable to the defence and unfavorable to the assailant; and when they do not exist naturally in the position, must be supplied by means of fortifications. These conditions for a position, and in consequence, for a fortification, may be briefly enumerated to be as follows:

1. A shelter must be provided to protect the defenders against the missiles of the assailant, and to screen them from his view.

2. The position should be so arranged that an assailant can not approach within cannon range and not be exposed to the fire of the defence.

3. If practicable, the position should be taken, and the shelter arranged, so that an approach of the enemy would be difficult, and the enemy's movements towards the position be greatly impeded.

4. The position and shelter should be arranged so that the defenders movements to defend the position should not be hindered or impeded in any way.

8. General method of complying with

these conditions.—The *first* condition or principle named is usually satisfied by making an excavation, and heaping the mass of earth thus obtained, until it is high enough and thick enough to screen and protect the defenders.

The *second* is fulfilled ordinarily by clearing away the ground in front of the position, and within cannon range, of all trees, houses, enclosures, hedges, etc., which might be used by the enemy as a screen; and by arranging the mass of earth used as a shelter, so that the defenders can fire over it, or through it, and sweep, with their fire, the ground over which the enemy must approach.

The *third* is observed by placing the fortification behind some natural obstruction, as a marsh, a water-course, etc. If this is not practicable, it is fulfilled by placing obstacles in front of the fortification and arranging them so that they shall impede the assailant's approach, but not screen his movements, nor protect him from the fire of the defence.

The *fourth* is complied with by removing any obstructions which might be in the way of the free movements of the defence, and by making communications which can be used by the defenders in moving from one part of the position to another.

9. Active and passive defence.—The terms, **active** and **passive**, are used to designate the kind of resistance which is offered in the defence of a position. If the defenders, seeing an opportune moment, should leave the position and attack the enemy without waiting to be attacked; or if during the assault, they leave the position for the same purpose, the resistance made is termed an **active defence.**

If no attempt is made to leave the work to attack the enemy, the resistance offered is known as a **passive** defence.

An active defence can be made, as a rule, only by a force which is strong enough and numerous enough to leave the position and become an attacking force. A passive defence is usually the resort of a force which is decidedly inferior to the assailant; or is the defence made when the object of the resistance is fully attained if the enemy fails in his assault.

An active defence requires that the third principle just mentioned should be modified. It requires that the arrangements, organized to impede an enemy's approach, should be made so as not to interfere with the free movements of the defenders.

10. Definitions.—The principal parts obtained in the construction of an earthern shelter are the mass of earth which is heaped up, and the excavation from which it was taken.

When the excavation is between the mass of earth and the enemy, it is called **a ditch**; if the mass of earth is between the excavation and the enemy, the excavation is then designated as **a trench.**

It will be seen that the trench is the simplest form of shelter, and the most quickly made. It fails to fully satisfy the conditions just named, as will be hereafter shown.

11. The ground occupied by a fortification is called the **site** of the work; and a plane tangent to this ground is called **the plane of site.**

The outline of a fortification as drawn upon the ground, or its projection upon a horizontal plane, is termed the **trace.** A section of the fortification made by a vertical plane passed perpendicularly to the principal or directing line of the trace, is called the **profile.**

12. Any position strengthened by fortifications made of earth is said to be **intrenched**; the fortifications used for the purpose are termed **intrenchments.** This latter term is usually applied to works of considerable extent which are used to shelter large bodies of men.

CHAPTER II.

ELEMENTS OF THE ORDINARY PROFILE.

13. Parapet.—Different names are used to designate the mass of earth employed for shelter, according to the object it serves.

When this mass is arranged so that the men using it as a shelter are enabled to fire over it and sweep the ground in their front with their fire, it receives the name of **parapet,** a word derived from the Italian [*para, petto,* defending the chest.]

If the mass is not arranged for this purpose, but simply used as a shelter, it is called an **epaulement.**

14. Terreplein.—The word, **terreplein,** is used to designate the surface on which the men stand in readiness to defend the parapet, and at the same time are screened from the enemy's view. The terreplein may be the natural surface of the ground, it may be above this surface, or it may be below it, as the bottom of a trench. In ordinary field fortifications the terreplein is the natural surface of the ground.

When the terreplein is above the natural surface of the ground, the latter is termed the **parade.**

15. Profile.—The form and dimensions of the **profile** of the parapet of an ordinary field fortification are represented in Figs. 1 and 2. The section ABCDEF is the profile of the parapet; the section GHJK is the profile of the ditch; and both together constitute the profile of the fortification, or as it is generally called, **the profile.**

This mass of earth must be made high enough to screen the men on the terreplein, A P, from the enemy's view, and it must be made thick enough to intercept his missiles, and thus afford shelter to those behind it.

When this mass of earth has a height, D D′, above the ground, of over four feet and six inches, some arrangement must be made by means of which the defenders can deliver their fire over it and upon the ground in front of it.

This is provided for in ordinary field fortifications, by means of a small terrace, B C, of earth, called a **banquette,** placed at a convenient distance below the top of the mass.

The upper surface of this terrace is called the **banquette tread** and is connected with the terreplein by a gentle slope, A B, called the **banquette slope.** Sometimes steps are used instead of the slope.

Even when raised high enough, the soldier would find it inconvenient to fire over the mass of earth, if it be left in the condition it takes when first heaped up. The upper surface of the mass is therefore sloped off at a convenient inclination, and joined to the ban-

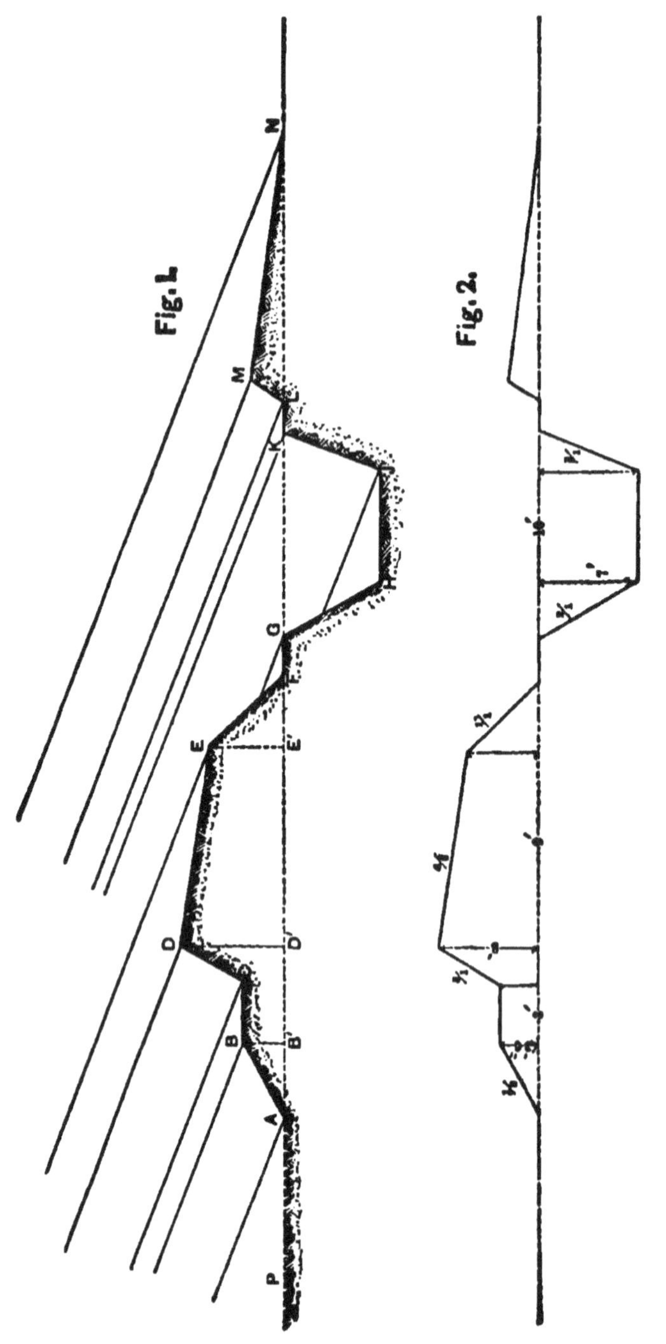
Fig. 1
Fig. 2
N
M
K
G
E
E'
D
D'
B
B'
A
P

quette tread by a slope, against which the soldier car lean in an easy position.

This upper surface, or top of the mass, is termed the **superior slope,** and the surface connecting it with the banquette tread is called the **interior slope.** The surface connecting the superior slope with the ground in front is called the **exterior slope.**

16. Berm.—The horizontal surface, F G, which connects the exterior slope of the parapet with the ditch is called the **berm.** The berm coincides with the natural surface of the ground in ordinary field fortifications.

17. Ditch.—The ditch may have almost any form of cross-section. The usual forms are triangular or trapezoidal.

The surface of the side next to the parapet is called the **scarp**; the surface opposite to the scarp is called the **counter-scarp.**

Sometimes there is placed a mass of earth, L M N, on the opposite side of the ditch, with the upper surface, M N, arranged with a gentle slope to the front. This gently sloping surface is called a **glacis,** a name also applied to the entire mass of earth so arranged.

18. Crest, foot, etc.—Particular names are given to the lines of intersection of the surfaces which have just been mentioned to distinguish them from each other.

The intersection of surfaces making a salient angle with the plane of site is termed a **crest**; if a re-entering angle, it is called a **foot**; and each intersection receives a characteristic name from the slope to which it belongs. Thus, the intersection of the banquette slope with the terreplein is, from the definition, a foot; and with the tread, a crest. These lines are repectively called "the foot of the banquette slope," and "the crest of the banquette slope."

The intersection of the interior slope with the superior slope is simply called the **interior crest**; while the intersection of the superior slope with the exterior slope is simply called the **exterior crest**; all others following the rule just given.

19. Principal lines of the profile.—The principal lines cut from the slopes by the vertical plane may now be enumerated. They are as follows, (Fig. 1):

The site, P G K N.
The banquette slope, A B.
The banquette tread, B C.
The interior slope, C D.
The superior slope, D E.
The exterior slope, E F.
The berm, F G.
The scarp, G H.
The bottom of the ditch, H J.
The counterscarp, J K.
The interior slope of the glacis, L M, and
The glacis, M N.

These lines which receive the names of the surfaces from which they are cut, intersect in points common to lines of the parapet and ditch, which points are known as follows:

The foot of the banquette slope, A.
The crest of the banquette slope, B.
The foot of the interior slope, C.
The **Interior Crest**, D.
The exterior crest, E.
The foot of the exterior slope, F.
The crest of the scarp, G.
The foot of the scarp, H.
The foot of the counterscarp, J.
The crest of the counterscarp, K.
The crest of the glacis, M.

CHAPTER III.

DIMENSIONS AND INCLINATIONS OF THE LINES AND SLOPES OF AN ORDINARY PARAPET.

20. Height of parapet.—The parapet must have a height sufficiently great to screen its defenders from the enemy's view.

A man standing on a level piece of ground when in the act of aiming his gun at an object, does not fire at a higher level than five feet above the ground on which he stands. A parapet five feet high would conceal from his view, when aiming his piece, all things behind it which do not rise above this height. I' the height is less than five feet, the parapet woulc only conceal the things which are no higher than it, and then only by placing them close to the parapet.

Taking the average height of the tallest men at six feet, the parapet should have a height of at least six feet and six inches to conceal men of this stature, when walking about on the terreplein.

From these two conditions the following principle has been stated, for determining the height of the parapet for an ordinary field fortification. It is as follows: *The height* of the interior crest of an ordinary field fortification *should not be less than six feet and six inches above the site,* and not less than five feet above the ground occupied by the enemy.

The labor required to construct a parapet increases rapidly with the height. A height of twelve feet is considered to be the greatest which it is practicable to give to the parapet of an ordinary field fortification.

The limits are, then, six and one-half feet, and twelve feet, for ordinary fortifications.

The height usually taken is eight feet, and this is the height of the parapet assumed for all field fortifications, unless otherwise stated, (Fig, 2.)

The settling of the earth after the parapet has been built; the wearing action of the weather; the possible effect produced by the enemy's projectiles; the amount of plunge in the trajectories of those missiles which graze the interior crests; etc, are all things to be considered, and afford good reasons for selecting eight feet as the least height to be used, under ordinary circumstances.

21. Command and relief.—The height of the interior crest above the site is the **command** of the work; its height above the foot of the scarp is the **relief.**

The term, command, is also used to express the height of the interior crest of one work above that of another; or above any particular point within range.

22. Thickness of parapet.—Sufficient thickness

must be given to a parapet to protect those behind it against the enemy's missiles. The thickness of a parapet is the horizontal distance between the interior and exterior crests. This of course must be greater than the penetration of the enemy's projectiles.

The kind of earth used in building the parapet, the penetration of the enemy's projectiles, and the probable duration of the enemy's attack, must be considered in determining the thickness to be given to the parapet.

The rule is to make the thickness of the parapet *one-half greater than the penetration* of the projectile into the same kind of earth as that of which the parapet is made. The amount of penetration will be different for different earths and will vary with the range, the calibre of the guns, and with the kind of projectile used. It is determined by experiment.

Two feet is about the limit of penetration of the the bullet, fired from the rifled musket, at close range, into ordinary earth. A thickness of three feet will give protection against musketry fire.

Six feet is the limit of penetration into ordinary earth for the projectiles of field guns. Nine feet would be the thickness to give to a parapet intended for protection against field artillery. As this is the usual fire to which ordinary field fortifications will be exposed, this thickness of nine feet is taken to be that given to the parapet, (Fig. 2) unless otherwise expressed.

It is well to remember that it is far better to make the parapet too thick than not thick enough. Thus, in the siege of Sebastopol, in 1854, the French did not make, in the beginning of the siege, a proper allowance for the increase of penetration of the enemy's projectiles, due to the large calibres of the guns used by the Russians. The result was that the siege batteries and powder magazines of the French were destroyed in the first bombardment. A similar incident occurred in the siege of Vicksburg, in 1863, where a battery constructed by the besiegers after several days hard labor, was in a short time knocked in pieces by the heavy projectiles of the besieged.

23. Banquette.—The banquette is a device by which the men are able to deliver their fire over the parapet. It is made just high enough above the terreplein to allow men of medium stature to fire over the interior crest. The distance of the tread below the crest is taken, for this purpose, at four feet and six inches; sometimes it is taken three inches less, or four feet and a quarter. The width of the tread depends upon the number of ranks expected to occupy it.

In the days of smooth-bores and muzzle-loading muskets, it was made wide enough for two ranks. It is rarely occupied, at the present time, by more than one rank.

A width of two feet is sufficient for one rank, although it is usually made *three* feet wide in ordinary field fortifications.

The tread is made with a slope to the rear, to allow the water falling on it to drain off. It is connected with the terreplein either by a slope or by steps. The inclination of the former is usually $\frac{1}{2}$; it may be greater if the banquette is low.

When steps are used, the tread of each step should not be less than twelve, nor more than eighteen, inches; the rise should not be less than nine, nor more than twelve, inches.

Steps are generally used whenever it is a matter of importance to gain space. All other things being equal, the ramp or inclined slope is preferred to steps.

24. Interior slope.—It would be desirable to make the interior slope vertical [and it is oftentimes made so] for the reason that the defenders in that case can stand close to the parapet in delivering their fire. A vertical slope would require a strong construction of some kind, to retain the earth in position, and to resist the horizontal thrust produced by the prism of rupture. When the materials for making this construction are abundant and convenient, a vertical slope, or one nearly so, may be used.

Under the supposition that such materials are not abundant nor convenient, it is usual to give an inclination to the slope, which, while convenient for the men to lean against, will require only a slight protection to preserve it.

This inclination is usually taken at $\frac{3}{1}$; an inclination which nearly coincides with that of the plane of rupture of ordinary earths. Earth having this inclination, and in a dry state, will exert little or no horizontal thrust. To preserve this inclination, it will only be necessary to protect it against the weather, or to use a protection strong enough to hold it up, when saturated with water. This protection or covering is made by facing the slope with a layer of stones, of wood, of sods, or any suitable material that will shed water, or be strong enough to resist the pressure, when the earth becomes saturated. This facing is called a **revetment**.

A steep slope requires a strong revetment, otherwise it is to be preferred. A gentler slope requires a slighter protection, but has the disadvantage of placing the soldier too far from the interior crest when he is in a standing position ; and it exposes him more to projectiles grazing the interior crest.

The rarity of hand-to-hand conflicts on the parapet, and the use of breech-loading weapons, allow the use of gentler slopes for the interior of the work than were formerly regarded as admissible. Gentle slopes are accompanied, however, by the disadvantage of requiring the soldier to occupy a recumbent position when firing his piece.

25. Superior slope.—The upper surface of the parapet is arranged so that a soldier behind the interior crest can reach with his fire a point near the crest of the counterscarp. This line of fire should pass within three feet of the crest of the counterscarp, and need not go below it. By this arrangement, the soldier can sweep with his fire all the ground beyond the ditch. By giving the superior slope an inclination of $\frac{1}{6}$ (Fig. 2.) it will be found that this condition will be fulfilled in an ordinary field fortification. If the line of fire along this slope does pass more than three feet above the crest of the counterscarp, the inclination must be increased. The angle must be limited, because as it is increased, the weakness of the parapet near the interior crest is increased. The limit of this inclination is $\frac{1}{4}$.

If a slope of $\frac{1}{4}$ does not bring the line of fire within three feet of the counterscarp, the inclination can not be increased. It is better in this case to construct a slight **glacis**, so as to bring the assailants under the fire of the parapet. Care must be taken that the upper surface of the glacis is kept at least five feet below the interior crest of the work.

It is well to state that a slope for the upper surface of the parapet is a necessary evil. It weakens the parapet by making it thin near the interior crest, and this thin portion wears away quickly under the action of the weather and the enemy's fire. Still, if it had no slope, it would mask the enemy when near the work.

26. Exterior slope.—From the outer edge of the superior slope the earth takes its natural slope. This is supposed to be at an angle of about forty-five degrees with the horizontal. The inclination of the exterior slope is therefore assumed to be $\frac{1}{1}$ (Fig. 2) unless otherwise stated.

In practice it is recommended to make this slope about $\frac{4}{5}$, or even $\frac{2}{3}$, since it is exposed directly to the fire of the enemy, and receives the drainage from the superior slope.

27. Berm.—The natural surface of the ground between the foot of the exterior slope and the crest of the scarp forms the **berm**, in field fortifications. Its width depends upon various circumstances, but principally upon the character of the soil. It is recommended to place the foot of the scarp upon the prolongation of the exterior slope. The width of the berm will therefore depend upon the inclination given to the scarp and the depth of the ditch. The steeper the slope of the scarp, the wider will be the berm.

As a matter of defence, the berm is considered to be a defect, since it offers advantages to the assailant in assaulting the work.

It is nevertheless a useful part during the construction of a work, as it forms a bench upon which the earth from the ditch can be thrown, and upon which workmen can stand to throw this earth upon the parapet. It is also useful in reducing the wearing effect of the wash upon the scarp, due to the water running off the superior slope, and down the exterior slope.

28. Ditch.—The primary object of the ditch is to furnish enough earth to build the parapet. The height and thickness of the parapet having been assumed, the cubical contents are easily calculated, and the dimensions of the ditch determined.

It is a principle belonging to the art of fortifica-ion, that everything used in the defence of a position should, when practicable, be arranged so as to be more or less of an obstruction to the advance of the enemy. The ditch should therefore be arranged so as to be an obstruction to the enemy's assault.

A ditch, not less than six feet deep and twelve feet wide, is an obstruction not easily passed. A ditch of less dimensions would not offer much of an obstruction to an assaulting column.

Ditches deeper than twelve feet are rarely constructed, in consequence of the amount of labor required to dig them.

Within these limits of twelve and six feet deep, a ditch can be dug which will give the necessary amount of earth and be at least twelve feet wide, thus offering a respectable obstacle to an enemy's assault.

In small ditches, it is recommended to make them

with a triangular cross section, in this way getting the greatest possible depth and width for the ditch.

In the profile (Fig. 2.) the ditch is trapezoidal in cross-section, with a depth of seven feet, and a width at bottom of ten feet.

29. Scarp and Counter-scarp.—The width at the top of the ditch is affected more or less by the inclination given to the sides of the ditch. The slopes of the counterscarp and scarp vary between wide limits, and depend upon the kind of soil, upon the depth of the ditch, the amount of obstruction the ditch is expected to give, etc. All things being equal, the slopes are made as steep as they will stand; and if required to last for some time, they should be protected by a revetment. In the ordinary profile (Fig. 2) they are taken at $\frac{1}{1}$ for the scarp, and $\frac{3}{1}$ for the counterscarp.

30. Calculation used to determine the dimensions of the ditch.

The height and thickness of the parapet having been assumed, the dimensions of the ditch may be obtained by the following method, which is practically that given in the "Aide-Mémoire," for engineers of the French army.

Denote by

R, the volume of the parapet.

S, the area of the profile of the parapet, and

l, the right line generated by the centre of gravity

of the profile of the parapet, supposing this profile moving parallel to itself, and generating the volume of the parapet under consideration.

Denote by

R', S', and l', similar quantities for the ditch.

The volumes for the parapet, and the ditch, for any part of the work under consideration, will be expressed as follows,

$$R = S \times l, \quad \text{and} \quad R' = S' \times l' \quad . \quad . \quad . \quad (1)$$

Earth when made into an embankment occupies a greater space than it did in the natural state. Denote this increase of volume by $\frac{1}{m}$. Since the volume of the earth in embankment is furnished by the volume excavated from the ditch, there results,

$$R = R' \left(\frac{m+1}{m}\right) \quad . \quad . \quad . \quad . \quad . \quad . \quad . \quad (2)$$

Substituting in equation (2) the values taken from eqs. (1), there results,

$$S' = S \, \frac{l}{l'} \left(\frac{m}{m+1}\right) \quad . \quad . \quad . \quad . \quad . \quad . \quad . \quad (3)$$

It will be sufficiently exact to take l' equal to the length of the middle line of the ditch ; which being substituted, gives S' in known terms.

Assume the slope of the scarp $\frac{1}{3}$, and, the counterscarp $\frac{1}{2}$ greater than the natural slope. Represent the width of the ditch at top by x, and its depth at the middle by y. Denote the angle of the natural slope by ϕ. Using this notation, the area of the profile of the ditch, is given as follows,

$$S' = y\left(x - \frac{7}{12}\,y \cot.\,\phi\right) \quad . \quad . \quad . \quad . \quad . \quad . \quad . \quad (4)$$

Solving eq. (4) with respect to x, there results,

$$x = \frac{7}{12}\,y \cot.\,\phi + \frac{S}{y}\,; \quad . \quad . \quad . \quad . \quad . \quad . \quad . \quad (5)$$

and solving with respect to y, and taking the minus sign of the radical, it gives

$$y = \frac{6}{7}\tan.\,\phi\left(x - \sqrt{x^2 - \tfrac{7}{3}\,S'\cot.\,\phi}\right). \quad . \quad . \quad . \quad (6)$$

From these equations (5 and 6) y can be assumed, and x deduced; or x assumed, and y deduced.

Making $\phi = 45°$, these last equations reduce to

$$x = \frac{7}{12}\,y + \frac{S}{y}, \quad \text{and} \quad y = \frac{6}{7}\left(x - \sqrt{x^2 - \tfrac{7}{3}\,S}\right)$$

It should be remembered in assuming values for x and y, that x must not be less than twelve feet, and y not less than six, nor greater than twelve feet.

31. In practice, it will be sufficiently accurate to calculate the area of the assumed profile, assume a depth for the ditch, and, without making an allowance for the increase of volume of the earth in the embankment, divide the area of the profile by the assumed depth of the ditch. The result will give the width of the ditch at the top.

32. Normal profile.—The profile (Fig. 2) with the dimensions and inclinations just mentioned is sometimes called the **normal profile** of field fortifications. It is the profile which would be constructed for a work located upon a level site, and when there is time to build it.

It is evident that great variations must occur, influenced largely by the kind of earth and the surrounding circumstances at the time of construction.

Slopes which are practicable in one kind of earth will not retain their shapes in other kinds.

Parapets placed on sites commanding all ground in common range need not be so high as those on lower ground commanded by neighboring heights.

The principles laid down and the reasons expressed for the normal profile apply equally well to all its variations.

CHAPTER IV.

THE TRACE OF A FIELD FORTIFICATION.

33. Trace.—The term, **trace,** is used by military engineers to denote the plan, or the general outline, of a field fortification upon the ground. (Art. 11.)

In field fortifications the governing or principal line used constructing the trace is the projection upon the plane of site, of the line from which the fire is delivered, viz, the interior crest.

The projection of the interior crest upon the ground is called the **sub-crest.**

34. Kinds of fire.—The trace of a field fortification can not be marked intelligently upon the ground, until the positions which the enemy may possibly occupy, and the kinds of fire he can bring to bear, are known.

Different names are used to designate the fire, both of artillery and musketry, depending upon its direction and kind.

The terms, **front, reverse, flank,** and **cross** fires are used to designate the direction of the fire with respect to the line aimed at. The projectile striking the line in front, or in rear, or at its extremities, or crossing in its flight other projectiles coming from a different direction, gives the name to the kind of fire to which the line is exposed.

When the direction of the fire is perpendicular, or nearly so, to the line aimed at, the fire is a **direct** one; if this direction makes an angle with the line aimed at, it is **oblique**; if this angle is very slight, it is a **slant** fire; if no angle is made but the direction coincides with the prolongation of the line aimed at, it is an **enfilading** fire; if it makes no angle, but is in front of the line, it is called a **flanking** fire; etc.

A line of troops or a line of parapet may be exposed to a front direct fire, a reverse fire, a slant fire, a slant reverse fire, an enfilading fire, a cross fire, etc. A column of troops might be exposed to a front direct, an oblique, a flank, a reverse, or a cross fire, according to the directions from which the projectiles came.

Other designations are used to denote the *kinds* of fire. Thus, a **direct or pitching** fire means one in which the projectile is fired from a gun at ordinary elevations, and with the service charge; a **curved** fire, when the angle of elevation is greater than usual, and the amount of charge used in the gun is less; a **vertical** fire, when the angles of elevation are still larger.

Other designations denote the kind of fire, as determined by the position of the line of fire with respect to the surface aimed at, or some peculiar char acteristic which marks it. Thus, a **ricochet** fire, is the result produced by small angles, low charges, and spherical shot; a **grazing** fire is when the projectile passes very near the surface; a **plunging** fire, when the projectile comes from a higher level than that occupied by the object struck; etc.

35. Salients and re-entrants.—An assailant when attacking a field work naturally tries to advance over that ground which offers the least obstruction to his free movements, and which does not expose him to the fire of the defenders. In the latter case, the most favorable ground will be that upon which the defence cannot bring a fire, and the most unfavorable will be that upon which the defence can bring a cross fire.

In order to have an effective direct fire upon the ground exterior to the parapet, the interior crest should be perpendicular, or nearly so, to the direction in which the fire is to be thrown. And since there are several directions in which the fire is wanted, it follows that the interior crest must be a **broken line.**

Those angles of the interior crest which project outwards, and towards the enemy, are called **salient** angles; those projecting inwards, are called **re-entering** angles.

36.—General principles. The following general principles should govern the engineer in the selection of the trace.

1. The trace should be as **simple** as possible.

2. The direction of the interior crest should be such as to admit of bringing **a strong direct fire** upon the ground liable to be occupied by the enemy.

3. The dimensions of the works and interior space should be **proportioned** to the number of men intended to defend them.

4. The principal lines of a fortification should be given such directions that they cannot be easily **enfiladed,** nor seen in reverse.

5. No salient angle should be **less** than 60°.

37. That the trace should be simple is evident. Field fortifications are works which are to be constructed in a short time and under pressing circumstances. Multiplicity of details and refinements of construction would cause a waste of valuable time, a thing of far greater importance in a campaign than theoretical perfection.

A strong direct fire is obtained, as previously stated, by placing the line of the interior crest as nearly perpendicular as possible to the direction in which the fire is to be delivered. A small deviation is admissible, depending upon the advantages to be derived from changing the position of the line.

The work should be proportioned so as not to require more men to defend it than are available. A fortification is a passive defence, and is an inert and helpless factor, if not defended by a live and active force.

It is not always practicable to arrange the lines so as to satisfy the second and the fourth conditions at the same time. To comply with the former would be, in many cases, to expose the line to the fire condemned by the fourth condition. The fourth condition must, under such circumstances, be satisfied by some expedient which will annul the effects of the enfilade fire. The expedients generally used are **traverses.** Their employment and construction will be explained hereafter.

Angles of less than 60°, for salients, restrict the interior space, and do not leave sufficient room to serve the guns which may be used in the salients; they also weaken the earth work of the parapet at these points.

It may not be possible to have the trace fulfil all these conditions. If it does, and if the profile of the parapet is a strong one, the fortification will be a good one and will prove a formidable obstacle in the enemy's way.

CHAPTER V.

FIELD WORKS.

38. Field works and lines.—Field fortifications are usually divided, according to the extent of the position fortified, into two general classes, viz.: **field works,** and **lines.**

The term, **field work,** is applied to the temporary fortification used to strengthen a position of limited extent, and in which the troops occupying it are expected to be dependent upon themselves alone for a successful resistance to an assault.

The term, **line,** is applied to the temporary fortification, or chain of fortifications, which is used to strengthen a position of considerable extent, and is to be defended by an army or a large body of troops.

These two classes of fortification do not differ as to the details of their construction.

39. Classes of field works.—The trace of a field work depends upon the directions in which it has to fire. (Art. 34.) It may have to fire on all sides of the position; or over a limited portion of a circle; or in special directions only, according to the positions which may be occupied by the enemy's artillery. These different circumstances give rise to three kinds of field works, viz.:

1. Field works exposed to the artillery fire of the enemy in one direction only, or in front;

2 Field works exposed to artillery fire of the enemy on the flanks as well as in front; and

3. Field works exposed to this fire upon all sides of the position.

It is evident that in the trace of a work of the first kind, the position which the enemy may have in its front need only be considered.

The trace of a work of the second kind must be considered under the supposition that the enemy may appear in front and on the flanks, that is in a portion bounded by a part of a circle.

The trace of a work of the third kind must be considered under the supposition that the enemy may appear on all sides of the work.

It is plain that parapets are needed on all sides, in the works of the third kind, to shelter the men from the enemy's fire; and that they are only needed on the exposed sides, in works of the first and second kinds. The works of the first and second kinds may then be left open on the sides not exposed to artillery fire, or they may be closed by some obstruction. From the construction of the parapets, in these different cases, there arise three classes of field works, viz.: **open, half-closed,** and **closed** works.

40. Open, and half-closed works.—A simple straight line of parapet is an example of an open field work. It is plain that a work of this kind would only be used where it would be impossible for the enemy to get on the flanks; or around in its rear; or where it is intended to abandon it as soon as the enemy gets on the flank or in the rear.

Half-closed works are used where there is any danger of the enemy appearing suddenly upon the flanks. They are closed by some obstruction so as to prevent a surprise or sudden attack by small bodies of infantry that might appear suddenly in rear of the position.

One of the simplest forms of the half-closed field work is the **redan**, (Fig. 3.) It consists of two straight lines forming an angle B A C, which is pointed towards the enemy. The angle at A is called the **salient**; the sides B A and C A are called the **faces**; the line B C is termed the **gorge**; the line A D is called the **capital**.

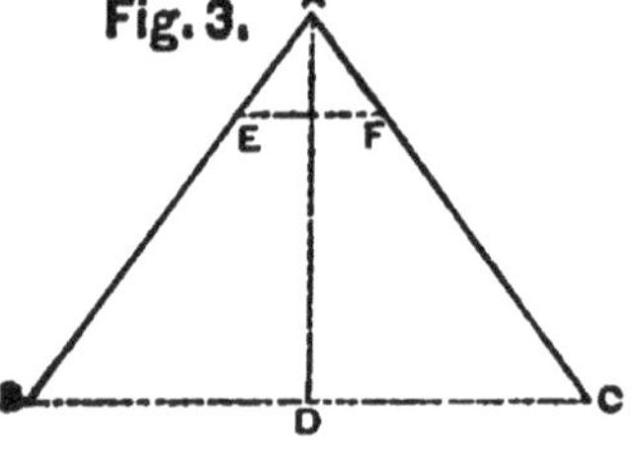

The sides of the redan are usually from thirty to sixty yards long. When the length is less than thirty yards it is called a **flèche.**

The redan, as here shown, delivers its fire over a part of a circle but has no front fire. If it be necessary to have a fire in the direction of the capital, it may be obtained by using the **blunted redan,** which is constructed by stopping the faces at points, as E and F, and connecting these points by a straight parapet. Two redans are sometimes placed side by side and joined to each other, making a work known as the **double redan**; sometimes the outer faces of the double redan are made much longer than the faces which are connected, in which case the work receives the name of **priest-cap,** or **swallow-tail.**

41. Lunettes.—If lines of parapet which are parallel, or nearly so, to the capital, are added to a redan, the construction will be that known as a **lunette**, (Fig. 4.)

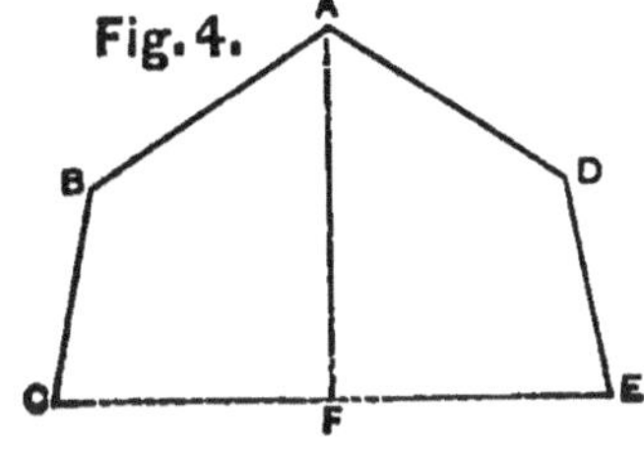

These parts, B C and D E which were added to the redan, B A D, are termed the **flanks**; the angles, at B and at D are called **shoulder-angles.** A lunette is therefore a field work consisting of two faces and two flanks.

42. Bastioned Front.—Suppose two lunettes,

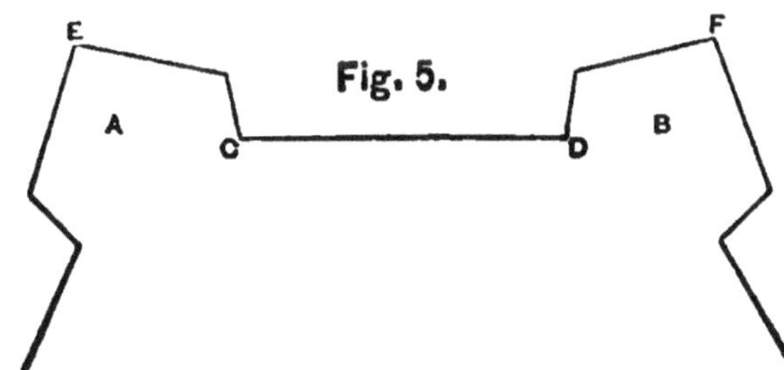

A and B to be connected by a straight line of parapet, C D (Fig. 5.) The resulting construction is called a

bastioned trace, and the portion between the capitals through E and F a **bastioned front**. The lunettes when thus joined are called **bastions**; the line of parapet joining them is the **curtain**; and the reëntering angles at C and D are known as the **curtain angles.**

43. Closed works.—Since the works of this kind are exposed to the enemy's fire of artillery on all sides, the position must be entirely surrounded by a parapet. The trace of such a work may be of any form, either circular, square, quadrilateral, polygonal—regular or irregular—all salient angles, or some salient and some reëntering angles.

A closed work in which all the angles are salients is called a **redoubt**; if there are reëntering angles, it is termed a **fort.**

44. Redoubts.—A redoubt may be of any figure provided it has no reëntering angles. Circular redoubts are sometimes used, but are objected to on account of the divergence of their fire. Redoubts are usually polygonal in plan, and, on horizontal sites, there is no reason why the plan should not be a regu-

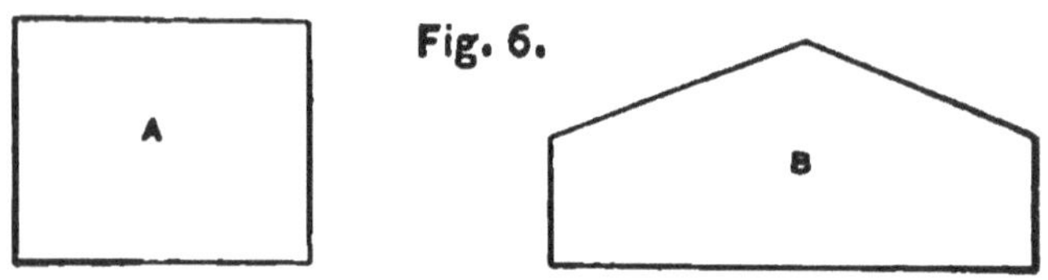

Fig. 6.

lar figure. The most simple and the most usually employed figure is the square, (A Fig. 6).

The redoubt, B (Fig. 6) is a form which is frequently used when it is desirable to combine its fire to the front with that coming from other points to the right and left of it, and still keep a direct fire on the flanks.

The advantages claimed for redoubts over forts are simplicity of trace, ease of adaptation to irregular sites, less labor required in their construction, and fewer men are necessary to defend them for the same amount of space enclosed.

45. Star forts.—A fort of this class receives its name from the general resemblance which its trace has to the conventional symbol used to represent a star. The star of six, or of eight, points is the kind mostly employed, although any number of points may be used, being determined ordinarily by the conformation of the ground on which it is built. The construction of the trace of a six pointed star is as follows:

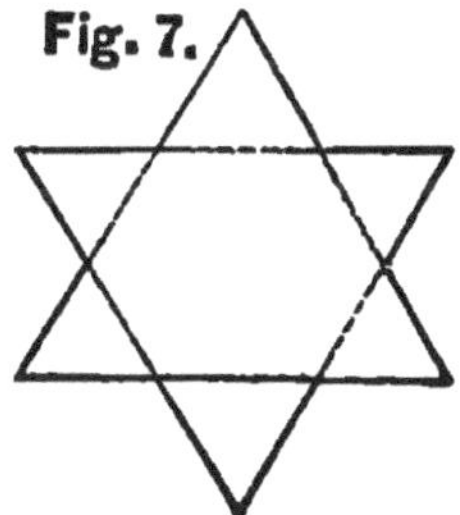
Fig. 7.

Describe an equilateral triangle; trisect its sides; on the middle portion of each side, determined by this trisection, construct an equilateral triangle. The resulting construction gives the six-pointed star, as shown by the full lines in Fig. 7.

The trace of the eight-pointed star is constructed in a similar manner, using a square instead of the equilateral triangle as the preliminary figure. The construction is as follows:

Describe a square; trisect its sides, and construct equilateral triangles on the middle portions thus determined (Fig. 8); or, what is a better trace, replace the right angles of the square by angles of 60°, as shown at A in the lower half of Fig. 8.

46. Bastioned fort.—If the space to be enclosed is circumscribed by a polygon, and on each of the sides a **bastioned front** (Fig. 5) is constructed, the resulting work is called a **bastioned fort.**

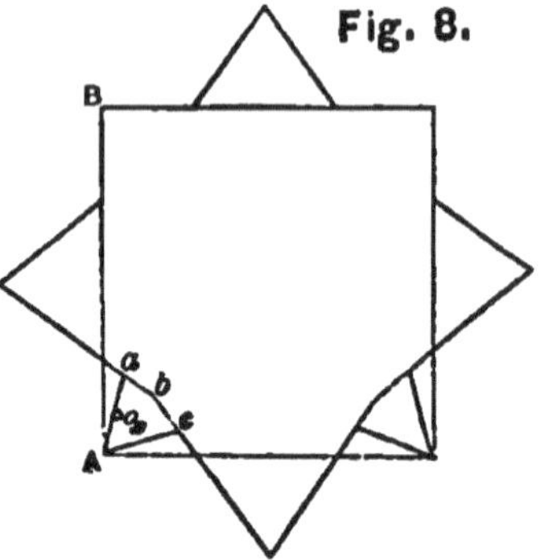

Fig. 8.

The polygon used to enclose the space to be defended may have any number of sides, and may be either a regular or an irregular figure. On a horizontal site, there is no reason why the figure may not be a regular polygon.

On a level site, the polygon generally used is a square unless a larger work is required.

The construction of the trace, when the circumscribing polygon is a square, is as follows:

Let A B (Fig. 9) be one of the sides of the circumscribing square. Bisect this side and at its middle point, C erect a perpendicular, C P and lay off a distance C P equal to one-eighth of A B. Join the point P thus determined with the ends, A and B, of the side of the square. Lay off from A and B,

Fig. 9.

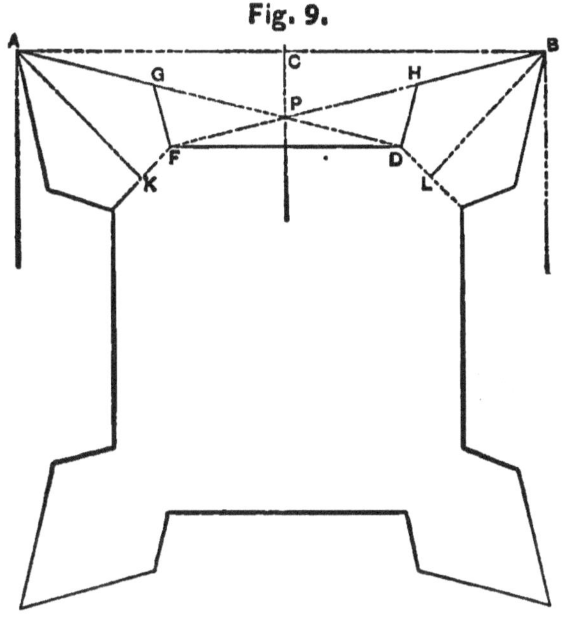

on the lines A P, and B P, the distances A G and B H, each equal to two-sevenths of A B. Draw through G the line G F, so that the angle G F B shall not be less than 90°, nor greater than 110°, and produce it until it intersects the line B P prolonged. Do the same at H, so that the angle H D A shall fall within these limits just named. Join the points F and D thus determined by a straight line. The line A G F D H B will be the trace of a bastioned front constructed upon the side A B of the circumscribing square. Do the same for the other three sides of the square, and the resulting construction is the trace of a bastioned fort.

In a similar manner the construction is made for a bastioned front, when the polygon has a greater number of sides than four. The only variation made is in the length of the distance C P, this distance being taken one-seventh of the side of the polygon instead of one-eighth, when the circumscribing polygon is a pentagon; and one-sixth, when the polygon is a hexagon or a polygon of a greater number of sides.

It is evident that the triangle can not be used for the circumscribing polygon. If it were, the fifth condition for field works (Art. 36) could not be satisfied.

47. Nomenclature.—The side of the circumscribing polygon is called the **exterior side**; the line, C P, is called the **perpendicular**; the lines D A and F B, the **lines of defence**; the angles G A C and H B C are called the **diminished angles**. The other parts are named as already stated in articles 40, 41, and 42.

48. Sectors without fire, and dead spaces.—The star and bastioned forts were devised to remedy the defects of sectors without fire, and of dead spaces, which are found in redoubts.

A **sector without fire** is the name given to that space exterior to a work which is not defended by the direct fire of the adjacent faces. Thus the faces A B and A C of the redan (Fig. 3) or the faces A B and A D of the lunette (Fig. 4) if prolonged, would include a space not swept by a direct fire from the adjacent faces. This space is in a great measure undefended, except by oblique fire, so far as the faces are concerned. The space included between lines drawn through a salient, perpendicular to the faces, is known as the sector without fire.

Any ground, over which the defenders' fire may pass, but so high above the assailant that he can not be injured by it, is called a **dead space**, or a **dead angle.** Thus, in the redan or the lunette just mentioned, if the enemy gets into the ditch, he is safe from any fire coming over the parapet. The ditch in each of these cases has the defect known as a dead space.

It was to remedy these defects, which exist in all works where the angles are all salients, that parts of a work were drawn back, or made to form re-entering angles, for the purpose of arranging lines from which direct fires could be brought to bear upon the ground not defended by the lines adjacent to it.

49. Flanked disposition. The arrangement of lines by which direct fires are brought to bear upon sectors without fire, and by means of which dead spaces are reached by a fire from the parapet, is called a **flanked disposition.** This term is used to denote this arrangement, because the enemy advancing upon a salient, where this arrangement is used, is exposed to a **flanking fire**—a fire parallel to and in front of the line attacked. (Art. 34.)

The defects of a flanked disposition are: the exposure of the lines of the work to enfilading or reverse fires; a contraction of the space enclosed by the work; and a partial sacrifice of the strong direct fire which the work might otherwise have.

A convergence or a crossing of fires upon the ground, which can be obtained by a flanked disposition at certain points, is productive of demoralization, as well as severe loss, among troops exposed to it. For this reason, a flanked disposition was formerly laid down as an essential element to be incorporated in all schemes of defenee. It is still an important factor. But, in the use of the long range weapons now employed, the strong direct and front fire is considered to be so important, that it is thought best not to sacrifice any part of it, if it can be avoided. Especially is this so when the advantage to be gained is accompanied by the defect of exposing some of the lines of the work to an enfilading fire.

An important advantage claimed for the flanked disposition is its ability to sweep its own ditches by fires from the work itself. In field works, the ditches are too narrow, and their depth too slight, to make this a matter of great importance, when compared with the defects which accompany the flanked disposition

Still, in field works, especially in the case of a large independent work, or a work which occupies the key point of an important position, the ditches should not be left undefended. They may be flanked by independent defences placed in the ditch—an expedient which will be explained hereafter.

50. Relation between the parts.—An intimate relation exists between the different parts of a flanked disposition. These are so connected that a change in any one affects more or less all the others.

The flanks are arranged to cross their fires in front of the salients, and to remove the dead spaces in the ditches (Fig. 9). To remove the dead space in the ditch in front of the curtain, requires that the fires from the flanks should reach the bottom of the ditch at the middle point of the curtain. To cross the fires effectively in front of the salients, requires that the line of defence should not be longer than the close and effective range of the weapon used to arm the flanks.

It is plain that the length of the exterior side must depend directly upon the length of this line of defence, and that the length of the curtain, which must admit of the ditch being flanked, must depend upon the relief of the interior crest of the flanks, and the inclination of the superior slope.

The shorter the line of defence can be made, at the same time keeping the curtain long enough to allow its ditch to be swept by the fire of the flanks, the more effective will be the cross fires in front of the salients, the longer will the enemy be under this cross fire, and the greater will be the chances of repulsing any assault made by him.

51. Least and greatest exterior sides of a bastioned front.—Taking the height of parapet at eight feet and the depth of the ditch at six feet, this relief of fourteen feet will be the least used; taking the height at twelve feet, and the depth at twelve, the corresponding relief of twenty-four feet will be the greatest used.

Assuming the superior slope at $\frac{1}{6}$, the least length of curtain for the least relief is fifty-six yards; the least length for the greatest relief is ninety-six yards.

Using the construction given in Art. 46, it will be seen that for a curtain of fifty-six yards in length, the exterior side must be about 125 yards long; for the curtain of ninety-six yards, the exterior side must be two hundred and fifty yards long.

The least length of the exterior side will therefore be between one hundred and twenty-five, and two hundred and fifty yards, depending upon the relief of the work.

The greatest length of the exterior side depends upon the length given to the line of defence. If the weapon used to arm the flanks is the rifled musket—the weapon now used by infantry—its close and effective range determines the length of the line o.

defence. The limit of accurate aim for the ordinary soldier is about three hundred yards. Assuming this to be the length of the line of defence, the exterior side will be about four hundred yards long, and will be the greatest length ordinarily used.

52. Defects of a bastioned fort.—The bastioned fort can not be used upon an irregular site, without sacrificing some of its most important qualities. It requires, also, considerable time and labor to build it.

For these reasons it is rarely used as a field work, except when it is to be an independent work, and used to defend some important point.

One point in the trace of the bastioned fort requires particular attention, and that is the counterscarp of the ditch. If the counterscarp is kept parallel to the interior crest, there will be a dead space in the ditch near the shoulder angle, and in the ditches in front of the flanks, unless a modification is made either in the profile or in the width of the ditch.

The modifications usually made are of three kinds, and are as follows:

One is to widen the ditch in front of the curtain by removing all the earth included between the curtain, the flanks, and the counterscarps of the faces prolonged. This method removes the dead space entirely.

Another is to widen the ditch at the top only, by giving the counterscarp of the ditch in front of the flanks and on the prolongation of the faces, a slope which will expose the bottom of the ditch in front of the flanks to the fire from the flank opposite. This method removes the dead space.

A third way is to slope the counterscarp of the prolongation of the ditch in front of the faces, so that the bottom of this ditch in front of the shoulder angles shall be swept by the fire of the flanks. This method removes the dead space in the ditches of the faces, but does not remove those of the flanks.

The first method is the one used in permanent fortifications. The amount of time and labor required to make this modification forbids its use in field works.

The second method has the same defects, but not to so great a degree.

The third method, though partial in it effect, is the one generally adopted in field-fortifications.

53. Defects of the star fort.—Star forts are better adapted to irregular ground than the bastioned forts. Otherwise, they possess all the defects of the latter, without the complete flanking arrangements which characterize the bastioned system.

CHAPTER VI.

LINES

54. Classes of lines.—The field works known **as** lines are divided into classes, according to the object for which they are constructed; or, according to some peculiar arrangement of their parts, or other characteristic quality. Some of the most commonly known are the **lines of circumvallation; of countervallation; intrenched camps; single and multiple lines; continued lines; lines with intervals; etc.**

The classification of lines into continued lines and lines with intervals is the one which is used in this chapter.

55. Continued lines.—When the entire front to be defended is covered by a *continuous* line of parapet, the work is called a **continued line.** There are no openings in a continued line, except those made for the use of the defence.

56. Lines with intervals.—When the front to be defended is covered by a number of field works, scattered along this front, and placed near enough together, to sweep the intervals with their fire, the whole arrangement forms a disposition called a **line**

with intervals.—Field works placed so near to each other, that each one can bring its fire to bear upon the ground in front of those adjacent, are said to be in **defensive relations** with each other.

I. Continued Lines.

57. Kinds of continued lines.—The principal types of continued lines in use are the **straight line**, the **tenaille line**, the **redan**, the **indented**, and the **bastioned lines**, which are easily distinguished from each other by their traces.

58. Tenaille line. The trace of the tenaille line is formed by making the alternate angles salient and re-entering; the condition being imposed that the

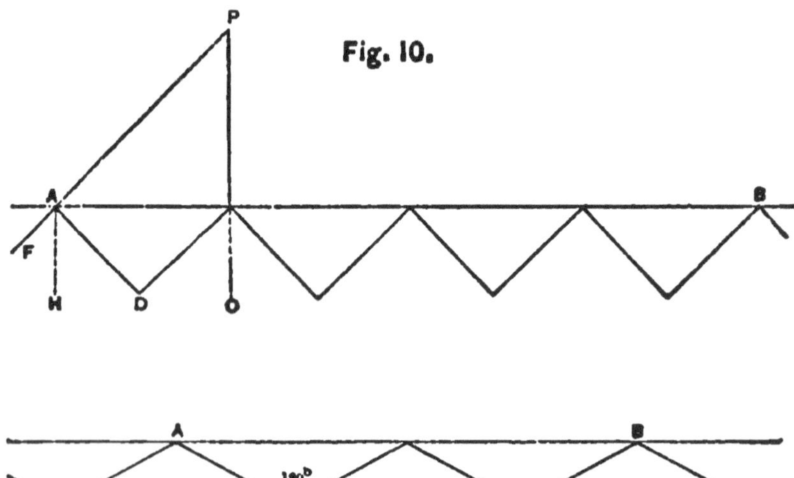

re-entering angles must not be less than 90°, nor greater than 120° (Fig. 10).

The line A B, joining the extreme right and left

salients, is called the **front**; and the ground in front of it, within range of the fire from the line, is called the **zone of defence**.

The greatest length of the faces depends upon the range of the weapon used in its defence. The main object of the tenaille line is to obtain a cross fire upon the zone of defence. This cross fire, to be effective, must be within the range of accurate aim of the weapon used.

To find the greatest length for the faces, the following construction may be used. Assume the angles for the salients, which must be known. Take the salient at A, and produce the face F A, until the distance A P, is equal in length to the range assumed. Through this point, P, draw a line P C parallel to the capital A H, and produce it until it intersects A B, the front of the line. Its point of intersection with this line will determine the position of the salient adjacent to A. Since the re-entering angle is not to be less than 90°, the greatest length of the faces, when the positions of the salients are fixed, will be when the re-entering angle is 90°. For angles greater the length of the faces will be less.

The least length of face is taken at thirty yards, so that there shall be no dead spaces in the ditch at and near the salients.

The more obtuse the salient angles are made, the more difficult it will be for the enemy to enfilade the faces. This exposure to enfilading fire, and the amount of labor required to construct the line, are the principal objections to its use.

59. Redan line.—A series of redans, placed at intervals and connected by straight lines of parapet, (Fig. 11) or by lines with very slight re-entrants, or salients, is termed a **redan line.**

Fig. 11.

The traces of the redans are usually made equilateral triangles. The salients are placed about two hundred and forty yards apart, and the faces are made sixty yards long. This line is sometimes known as **Vauban's line.**

There is another redan line composed of large and small redans, like that shown in Fig. 12. It is

Fig. 12.
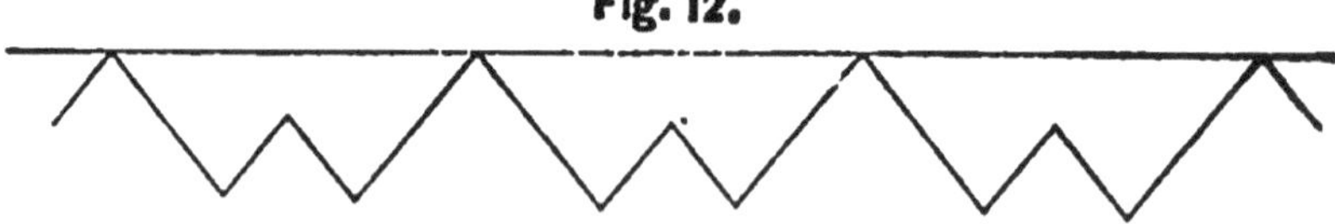

known as a **continued redan line**, to distinguish it from Vauban's line.

60. Indented lines.—The indented, or **crémaillère** line consists of short and long branches, which may be arranged as represented in Fig. 13.

The long branches are usually made seventy yards long, or more, and are directed towards ground which cannot be occupied by an enemy. The short branches are made about thirty yards long, and are used to

Fig. 13.

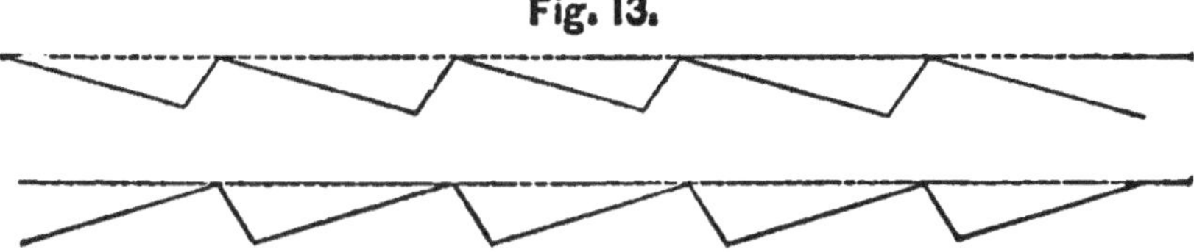

flank the long branches. Instead of giving the long branches directions parallel to each other, as indicated in Fig. 13, they may all be directed upon a single point, which the enemy cannot reach, as shown in Fig. 14.

Fig. 14.

61. Bastioned line. A bastioned line is composed of a series of bastioned fronts, joined to each other on the same general line, (Fig. 15) the salients

Fig. 15.

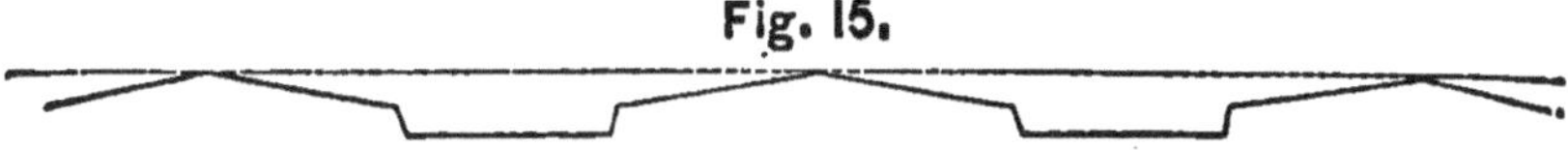

being placed from two hundred and fifty to four hundred yards apart. The line joining any two consecutive salients is taken as an exterior side, and a bastioned front constructed upon it by the rules already given.

62. Advantages and disadvantages of continued lines.—Continued lines have been much used in past military operations, and will, in all probability, be used in the operations of the future.

Continued lines may be usefully employed where a passive defence only is to be made, and where the position to be defended is limited in extent, and not exposed to flank attacks.

They are not fitted for an active defence, and they have the serious disadvantage of being untenable, when any part of the line has been taken by the enemy.

They require a large amount of labor to construct; and it is a very doubtful question, whether the advantages they give compensate for the time and labor employed in their construction.

II. Lines with Intervals.

63. Lines with intervals.—This class of lines differs from the continued lines, by leaving intervals along the front of the position, which intervals present no obstructions to an enemy moving through them, excepting so far as they may be defended by the fire of the works, or may be obstructed by natural obstacles, or by artificial ones placed along the front.

The works forming the line may be placed so close to each other, as to be in defensive relations; or they may be so far apart, as to admit only of their defending the intervals between them.

If the works forming the line are to be in defensive relations, that is, if they are to afford a reciprocal defence, it is evident that the remarks made upon obtaining the length of the line of defence, in a flanked disposition, apply equally to determining the distance between the works. Thus, a line of redoubts (Fig. 16) placed so that their salients are three hun-

Fig. 16.

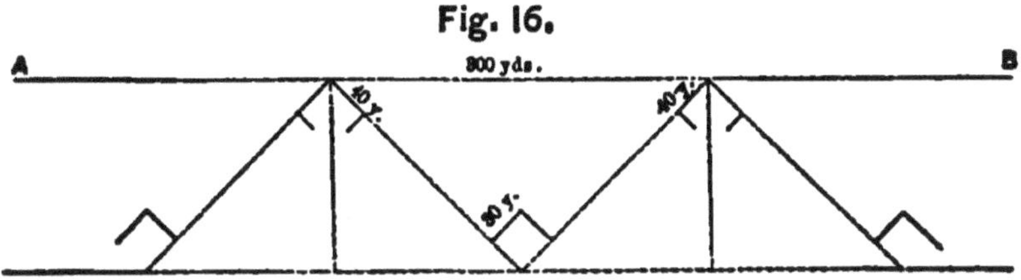

dred yards apart, along the line, A B, would have the works on this line in defensive relations with each other, the weapon used being the musket.

If it be required to defend only the intervals between them with the musketry fire, this distance between the salients of the consecutive works might be as much as six hundred yards, as this would bring the middle portion of the interval within the range of accurate aim.

If the defence is to be made with artillery, the distance, in the first case, might be one thousand, or even so great as one thousand five hundred, yards. In the second case, these distances may be doubled.

If an obstinate defence is to be made, a second line should be used (Fig. 16). A second line is especially useful when the works on the outer line are not in defensive relations with each other.

The second line should be placed behind the first, and distant from it, about one-fourth, and certainly not greater than one-half, of the distance between the works in the first line. When artillery is used in the second line, a good position would be about six hundred yards distant from the first. This places the second line just out of range of aimed musketry fire, but in close range of artillery fire.

64. A third line of field works is sometimes built. The general arrangement of the works of this third line, with the works of the first and second, conforms to the principles employed in arranging the works in the second line (Fig. 17).

Fig. 17.

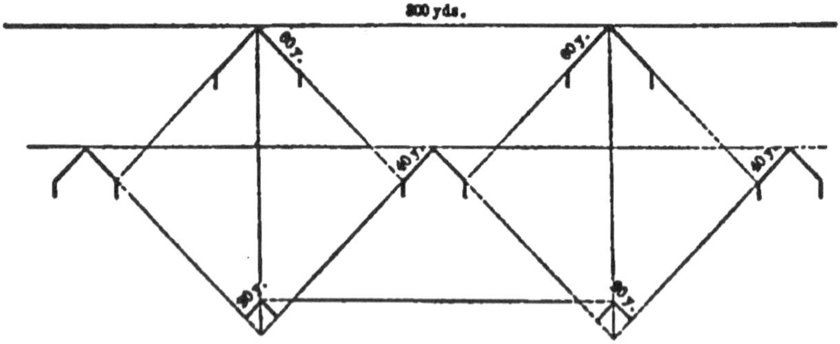

A third line might be useful, in case of an active defence, since the works placed along this line can be utilized as screens for the reserves and for bodies of cavalry.

A fourth line would evidently be of no practical service in the defence of a position. A force, unable to retain possession of the outer lines, could not be expected to hold the fourth.

The number of lines, whether a single one, or two, or three in number; the kinds of works to be used on each line; the distance apart of the works on each line; the distance apart of the lines; and the details of their construction, depend upon the natural features of the ground, the numbers and kinds of troops which are to occupy them, the range of the arm used in their defence, and the time disposable in which to construct them.

65. Advantages and defects of lines with intervals.—Certain advantages are claimed for lines with intervals. These advantages may be briefly stated to be as follows:

1. The lines with intervals admit of either passive or active defence. They are peculiarly fitted for the latter.

2. Lines with intervals are more easily adapted to the natural features of the ground than continued lines.

3. For the same extent of front, they require fewer men to defend them, and the works require less labor to construct, than other kinds of lines.

4. They admit of a better use being made of raw and inexperienced troops.

5. A line with intervals may still be defended, even after a part of the line has been captured, or after the enemy has broken through the line.

The main defect inherent in lines of this class, is the sub-division of the defenders into several independent commands, by which unity of action of the whole command is impaired.

This defect is reduced somewhat by a proper disposition of the works. A few capacious and strong works are better than a large number of small ones. Experience has shown that a body of one thousand men, in a single well-constructed work will offer a more effective resistance to the attacks of an enemy than the same number would, when scattered among three, four, or more, smaller works. The difficulty which a general would meet, in obtaining experienced officers fit for these independent commands, and in having these officers to act in unison with each other, gives sufficient grounds for such a result to be expected under ordinary circumstances.

CHAPTER VII.

THE SIZE OF A FIELD WORK, THE NUMBER OF THE GARRISON, AND THE SELECTION OF A TRACE.

66. Size.—The size of a field work, when built upon a level site, depends principally upon the number of men intended for its defence. When built upon an irregular site, the size, in addition, depends frequently upon the extent of ground to be swept by the fire of the work.

A good defence of a work is obtained only by a strong musketry fire, which is only obtained by allowing a musket for each pace, measured on the interior crest. The number of men to form a single rank and to furnish this fire would be equal to the number of paces contained in the length of the interior crest.

It will be sufficiently accurate to assume four paces for each three yards in length. An interior crest of three hundred yards would require four hundred men to line it in single rank.

A vigorous defence requires not only men enough to line the parapet, but a good many more to supply the vacancies from casualties or other causes, and to furnish support to the line. Double the number required to man the parapet is considered to be sufficient. That is, a vigorous defence of three hundred yards of interior crest would require a total of eight hundred men.

Hence, to determine what the length of interior crest shall be, the number of men being known, it will only be necessary to divide the number of men by two, and take this quotient for the number of paces which the interior crest should have.

Conversely, the length of interior crest being known, it will only be necessary to find the number of paces in its length, and double this number for the number of men which will be required.

67. Garrison.—A body of troops stationed in or near a field work, to defend it, is called its **garrison.**

The garrison of a work, when it is practicable, should always be a complete organization, or composed of detachments belonging to the same unit of force.

Garrisons should not live within field works, unless there is a pressing necessity for this to be done. As a rule, they should encamp near the works they are to defend, and only keep guards within the works.

Nevertheless the engineer, or other officer, who lays out a field work should always consider the possibility of its being occupied by a garrison, and should provide the necessary accommodations, so far as interior space may be required.

The amount of interior space to be enclosed becomes a matter of importance under these circumstances, and especially in redoubts.

68. Enclosed space.—The amount of space to be enclosed by a redoubt can be easily calculated.

It is usual to allow from one and a half to two square yards of space for each man; a space of seventy-five square yards for each field gun; a space of twenty square yards for a powder magazine; and a space of one hundred and seventy-five square yards for a traverse which is used to protect the outlet, or passage way, leading out of the work.

Suppose a field work is to be built, which is to be a square redoubt, to be armed by four field guns, and to have a traverse opposite the outlet; what is the area to be enclosed, and the length of interior crest, the garrison numbering four hundred men?

From the quantities above, the space for the accommodation of the men, etc., will be as follows:

400 men, allowed 2 square yards each,	800	square	yards.
4 field guns, " 75 " " "	300	"	"
1 powder magazine,	20	"	"
1 traverse,	175	"	"
Total	1295	"	"

In addition to this space, there must be room for the part of the parapet between the interior crest and the banquette slope. The distance from the foot of the banquette slope to the sub-crest, in the ordinary profile, is three yards. This distance must be enclosed, as necessary for the troops.

The square root of 1295 may be taken to be thirty-six.

A square whose sides are each thirty-six yards long will enclose a space sufficient for the accommodation of the men and guns, and allow for a traverse and a magazine. If a square be taken whose sides are six yards longer, that is, each side equal to forty-two yards in length, it will enclose a space sufficient for the men, guns, etc. ; and will also allow room for the parapet.

A square redoubt whose sides are forty-two yards ong, or whose entire interior crest is one hundred and sixty-eight yards long, will enclose space enough to accommodate a garrison of four hundred men, four guns, etc.

The rule given in Art. 66, would require a force of four hundred and forty-eight men to defend this work. It is evident, then, that there must be a least length of interior crest, which being defended according to the rules laid down, shall enclose the necessary space for the accomodation of the garrison.

69. Relation between least length of interior crest. and space enclosed for a square redoubt for a given number of men, etc.

Let x represent the number of yards in the interior crest of one of the sides of the redoubt;

n, the number of men forming the garrison;

s', the number of square yards allowed for the guns, magazine, and traverse.

Let it be supposed that each man is allowed two square yards, and that for every three yards of interior crest, there shall be eight men.

These conditions may be expressed by the following equations.

$$(x-6)^2 = 2n + s' \quad . \quad . \quad . \quad . \quad . \quad . \quad . \quad . \quad (1)$$

$$\tfrac{1}{8}n = \tfrac{1}{3}x \times 4 \quad . \quad . \quad . \quad . \quad . \quad . \quad . \quad . \quad (2)$$

If the redoubt is to be armed with field guns, and to have a powder magazine and a traverse. the value of s' may be determined by the rule already given.

Suppose the redoubt is to be armed with four field guns, etc., and that the space allowed for them is that named in the last article. After substituting for s', its value, four hundred and ninety-five square yards, equations (1) and (2) may then be combined and solved with respect to x and n. Their values will be found to be

$$x = 44, \qquad \text{and} \qquad n = 473.$$

If the redoubt is not to be armed with artillery, but is to have a traverse opposite the outlet, the value of s' will be one hundred and seventy-five yards, and the corresponding values for x and n, will be

$$x = 37, \quad \text{and} \quad n = 395.$$

All square redoubts in which the lengths of the sides are less than thirty-seven yards and forty-four yards, will not have the requisite space enclosed. Since the area increases in a much more rapid ratio than the perimeter, all square redoubts whose sides are longer than those named, will contain more space than is absolutely sufficient.

70. Selection of the trace. The selection of the trace of a field work is dependent upon the number of men that can be spared to defend it, and upon the time that can be allowed in which to build it.

The selection is also dependent upon the natural features of the site. Thus, suppose a field work is to be placed in front of a bridge, to defend the crossing of the river at that point (Figs. 18 and 19). The

Fig. 18. Fig. 19.

redan, which was a fitting trace where its site is in the bend of the river (Fig. 18) would not be so well fitted for the defence as a lunette would be, if the site is bounded by a straight course of river (Fig. 19.)

The selection is also influenced by the particular

object to be attained. It is plain that a trace which would bring a strong direct fire upon ground which an enemy would not occupy, would be a faulty one. The arrangement of the trace upon the site, after the trace has been chosen, is an important matter, and may be considered as a part of the selection.

Thus, a field work to be built in rear of a marsh, the crossing of which is impracticable for the enemy (Fig. 20) would have a faulty trace, if its lines were

Fig. 20.

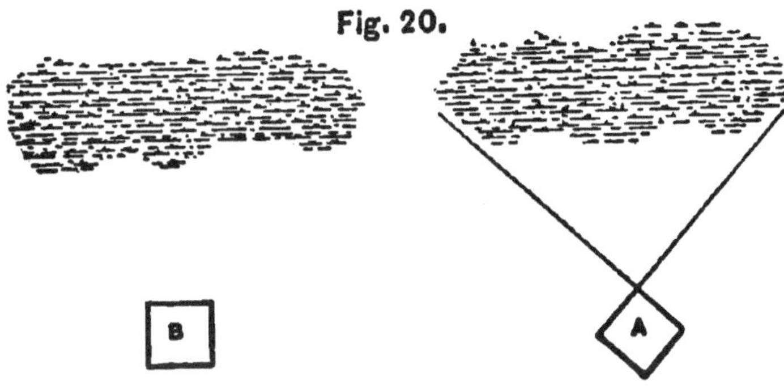

so placed as to bring strong direct fires upon the marshy ground, and not upon those portions along which the enemy would be obliged to approach in his advance upon the work.

The same fault would exist where the trace was assumed, if its lines were not properly directed. Thus, suppose a square redoubt was to be placed in rear of the marsh, it is plain that the redoubt B (Fig. 20) would have the defect just named, while the redoubt A would not.

If, instead of a marsh, the ground had been high,

it is easily seen that the redoubt at A is better placed than the one at B.

No absolute rules can be laid down which will apply to all cases. The engineer must exercise his skill and ingenuity in each case, taking care to harmonize as far as possible the conditions to be observed, whenever they conflict.

CHAPTER VIII.

CONSTRUCTION OF FIELD WORKS.

71. Tracing.—After having determined upon the dimensions of the profile, and having selected the trace, the first step in building the work is to mark its outline upon the ground. This operation is called **tracing**. The operation of tracing consists in marking the sub-crest and other necessary lines upon the ground, so that they can be distinctly seen, and can be used to determine all other lines and points of the work.

The "tracing" of a field work may be made by means of a drawing which represents the work to be constructed; or it may be made directly upon the ground without the use of a drawing.

When a drawing is used, it is usual to take two points upon this drawing, which can be easily located upon the ground, and join these points by a straight line. This line is then taken to represent a base line which can be used in laying out the work. From the different angular and important points of the plan, perpendicular lines are drawn to this base line. The distances to the base line, and the distances intercepted upon it, are measured and noted upon the drawing. Going to the site, the two assumed points are located upon the ground, and a straight line is drawn through them. This line is the base, and is marked upon the ground either by a cord, or by a furrow in the ground made with a pick. The distances are then measured off on the base line, and the points marked where the perpendiculars are to be constructed. These perpendiculars are then constructed by some of the simple methods used for this purpose, and the distances to the angular points laid off on these. These points, thus determined, are then marked by stout pickets driven into the ground. The angular points are then joined by straight lines, and these lines are then marked upon the ground by a cord stretched between the adjacent points, or by a furrow made in the ground. The marking of these lines completes the tracing of the sub-crest.

If there is no drawing, but the work is to be laid out directly upon the ground chosen, it is usual for the engineer to first select the salient points of the work, and mark them by stout pickets driven into the ground. He may then determine the reëntrants by inspection, or by some rule. Which method he will use will depend upon the kind of tracing to be made. The angular points, all determined, are then connected by straight lines, marked as in the previous case.

72. Profiling.—The operation called **profiling** forms the next step in the construction. This operation consists in erecting, at proper points along the sub-crest, wooden **profiles** which give the form of the parapet at those points, and which guide the workmen in the construction of the work.

Profiles are placed at the extremities of a parapet; at points along the sub-crest from twenty to thirty yards apart; at the salients and reëntrants; and at any points where a change is to be made in the dimensions, or in the direction of the parapet. The profiles are made perpendicular to the subcrest, excepting at the angles, where they are made to bisect the angles.

An ordinary profile may be constructed as follows: A straight line, as D′ K (Fig. 21.) is drawn on the

Fig. 21.

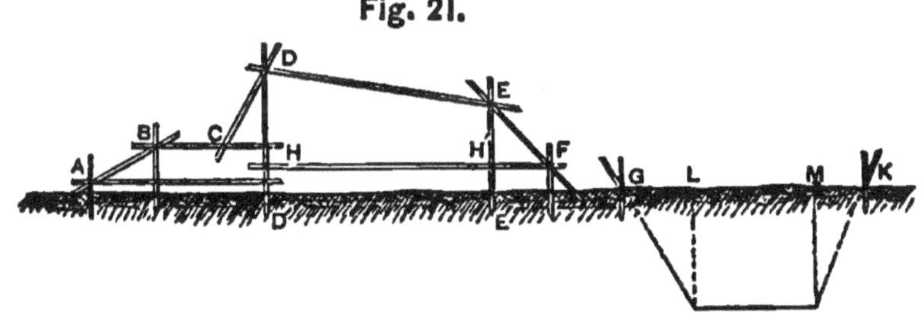

ground at D′ and perpendicular to the sub-crest. A distance, D′ E′, equal to the thickness of the parapet, is laid off and marked by stout pickets driven into the ground, at D′ and E′. Poles or strips are nailed to these pickets in a vertical position, and held firmly in place by a strip, H H′, fastened to them. It will be found convenient to make this strip horizontal, which may be done by a level.

The height of the interior crest is marked on the pole at D′, and the exterior crest on that at E′. A wooden strip, from two to three inches wide, and one inch thick, is then nailed to these poles at the points D and E, just marked. These wooden strips may be made by sawing boards into pieces of the necessary width.

A distance H′ F is laid off on the horizontal strip H H′, equal to the base of the slope which the earth is to take. A stake is driven at this point, and the point F marked upon it. A strip, E F, extended until it meets the ground, is then nailed to the uprights, which strip, in this position, marks the inclination of the exterior slope.

Similar strips are fastened to the other uprights to mark the inclination of the interior slope, the width of the banquette, and the inclination of the banquette slope. The profiles being finished, the foot of the banquette slope and the foot of the exterior slope are marked upon the ground in a way similar to that used for the sub-crest. This operation completes the profiling and tracing of the parapet.

The profile of the parapet being finished, the points G and K. or points of the scarp and counterscarp, are marked by pickets driven into the ground. The crests of the scarp and counterscarp are then traced upon the ground. Where attention is to be paid to the slopes of the ditch, short strips are nailed to the pickets, marking the inclinations to be given. In addition, the points L and M are determined, and the lines projected in these points are traced upon the ground by means of a pick.

73. Execution of the work.—The third step is begun by bringing the working parties on the ground and commencing the excavation.

The first thing to be done is to remove all brush, trees, rocks, or obstructions, which may hinder the work of excavation.

In this kind of work, as well as in the excavating, the working parties are divided into smaller bodies, generally termed **gangs.**

The gangs, being furnished with shovels and pick-axes, are distributed along the line of the ditch, under the supervison of competent foremen.

It is recommended to divide the crest of the counterscarp into lengths of six feet or more (sometimes twelve) and to mark the divisions by small pickets. The sub-crest is then divided into the same number of divisions, which are connected with those already marked upon the counterscarp, by furrows traced upon the ground, these furrows being prolonged until they intersect the foot of the banquette slope. In this way, there are formed small areas, in each of which a working party can be placed, and in each of which there will be room for the party to work without crowding. The number of workmen, the proportion of picks and shovels, and the positions of the men, will depend upon the nature of the ground to be excavated and the distances to which the earth must be thrown.

In ordinary soils, one man with a pick can loosen, in a given time, as much earth as two men can remove with shovels, during that time. The amount of earth removed can be determined experimentally, by having a man to loosen the earth over a given space to the depth of one blow of the pick, and note the time employed. Then have a man to shovel this loosened earth off, and throw it a horizontal distance of twelve feet, and note the time taken to do it. With the data thus obtained, the proportion of picks to shovels can be determined.

It is assumed that a workman while shovelling can pitch a shovelful of earth to a distance of twelve feet in a horizontal direction or of six feet in a vertical one.

The shovellers in the beginning of the excavation can therefore be placed twelve feet apart; but as the ditch deepens, the distance apart must be lessened, and finally, the number of shovellers must be increased.

There should be for every two or three working parties, one man, at least, to spread or level the earth as it is thrown upon the embankment, and another man to ram it.

From the foregoing, knowing the kind of soil and

the dimensions of the ditch and parapet, the size of each working party can be estimated.

In the field, the workmen are soldiers, taken from infantry regiments, detailed for this fatigue duty The details are divided into squads, forming working parties under the supervision of their non-commissioned officers. Several of these squads form a fatigue detail under the charge of a commissioned officer and are by him, divided into reliefs according to the circumstances of the case. The soldiers are distributed, as just described, along the line of the ditch, excavating the earth, throwing it upon the embankment, spreading it, and ramming it.

There should be present with every fatigue detail an engineer soldier, who should explain to the men their duties, before they begin to work; should advise them how to use their tools; and should be responsible that the proper slopes and dimensions of the profiles are observed.

74. Time required.—The day's task for an ordinary workman in common soil is about eight cubic yards, when the earth is not thrown higher than six feet. Half of this, or about **four** cubic yards, may be taken as a fair day's work for the ordinary soldier on fatigue duty, if he works faithfully. The time for throwing up a work, whose profile is that shown on page 17, may be roughly calculated, as follows:

The area of the section of the parapet is about

one hundred and eighteen square feet. The area laid off for a working party has a length of six feet. It may be roughly estimated that the working party assigned to one of these areas is required to make an embankment containing seven hundred and eight cubic feet, or about twenty-six cubic yards. An excavation which measures twenty-three cubic yards will give earth enough to make a parapet containing twenty-six cubic yards. Calling the amount of excavation to be twenty-four cubic yards, and supposing each party to take out only four cubic yards per day, it will require **six** days to construct the parapet.

The parapet, of the dimensions given, can be constructed by fatigue details, in one-third of this time, or less, by increasing the number of men in each working party, dividing the party into reliefs, and working continuously until the parapet is finished.

When the embankment has reached the height of the banquette tread, additional working parties are organized, for the purpose of constructing the revetments, which are required to support and protect the interior slope.

When no engineer soldiers can be spared to oversee the work, substitutes must be obtained. There will be found, in all of the infantry regiments of the United States service, non-commissioned officers who have had experience in the labors of excavating and embanking. These can be detailed to act as engineer soldiers and can fit themselves in a very short time to discharge the duties assigned to them.

The method of posting the men is a matter of detail, acquired from practice; the only condition imposed is that there shall be no *crowding.* Only a limited number of men can work at the same time, and rapidity of execution can only be obtained by frequent reliefs. Rapidity of execution is facilitated by good judgment, especially in the selection of the men. Good axemen should not be used to dig; and good diggers should not be used as axemen. Proportioning the work according to the skill displayed by the men will materially shorten the time required for its execution.

CHAPTER IX.

REVETMENTS.

75. Definition.—Any facing used to protect a slope of earth from the action of the weather is termed a **revetment.** (Art. 24.)

Embankments of earth are frequently made with slopes greater than the one which the earth naturally assumes. These slopes, when left exposed to the action of the weather, soon wear away and lessen until they coincide with the natural one. Where it is necessary to preserve them, it is done by using revetments, or retaining walls. The latter are necessary when there is a horizontal thrust due to a mass of earth, which would fall were it not supported. The former are used where there is little or no thrust, and the object is principally to protect the slopes from the weather.

76. Materials used for revetments.—If there is time, and the work is to be kept in good condition, it is a good policy to protect all the slopes, by using revetments, or by sowing the slopes with grass seed.

The interior slope is, however, the only one in a field work that requires protection as soon as constructed. All other slopes can wait until the parapet is completed. Field works are frequently constructed hurriedly, and, in consequence, the revetment of the interior slope must be made quickly. This necessitates the use of any materials suitable for the purpose, and which can be quickly procured. Some of the materials near at hand, or which can be quickly procured, from which the revetments can be made, are **fascines, gabions, hurdles, timber, plank, casks and barrels, sandbags, grass sods, etc.**

77. Fascines.—A **fascine** is a long, cylindrical, bundle of brush-wood, bound closely together near the ends and at intermediate points by withes or by wire (Fig. 22).

Fig. 22.

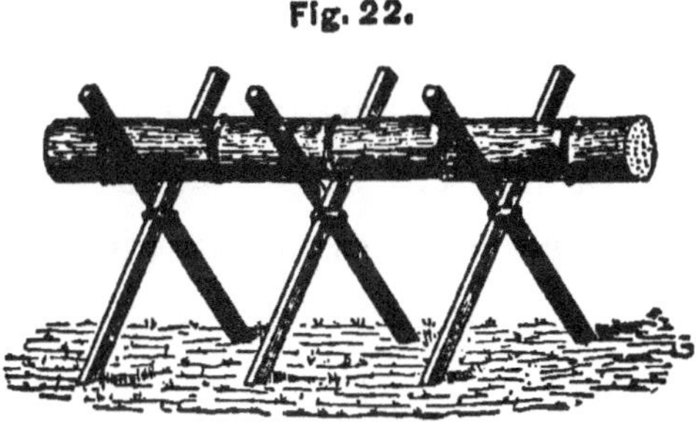

A fascine is usually made nine inches in diameter, and eighteen feet long. The method of making them is as follows: Trestles are placed in line, with the end ones sixteen feet apart from each other. If five trestles are used, the trestles would be four feet apart. Each trestle is made by driving two stakes, each six and a half feet long, and three inches in diameter, obliquely into the ground so that they will nearly make a right angle with each other, and binding them together at the middle. The brushwood, stripped of its leaves and small twigs, straight and flexible, and about an inch in diameter, is laid in the trestles, as nearly uniformly in thickness as possible, until the trestles are nearly full.

This brushwood is compressed into a compact bundle by means of a **fascine choker** (Fig. 23), and bound firmly by bands.

Fig. 23.

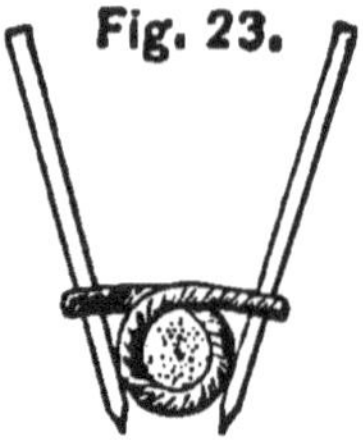

The fascine choker is formed of two stout levers, which are connected by a piece of chain or stout rope, as shown in the figure. Its action is easily understood. The bands are usually **withes**, sometimes called **gads**, prepared for the purpose. The bands are twelve in number, placed at equal distances apart; those near the ends being six inches from the end of the fascine. Wire, and even spun yarn, are used for the bands at times.

Every third or fourth band should be made with a projecting end, terminating in a loop. This loop is used to go around a small picket driven into the earth, and hold the fascine in its place.

After practice, a fascine can be made by a squad of five men in about an hour.

78. Fascine revetments.—The revetment of

the interior slope, by using fascines, is shown in Fig. 24.

Fig. 24.

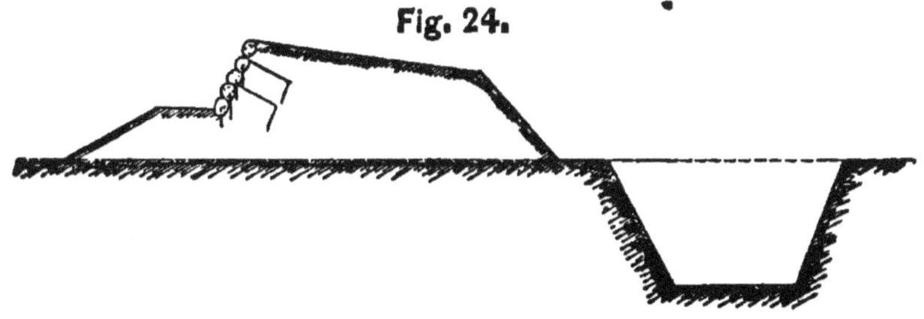

The first row of fascines is partly imbedded in the earth, so that about half of the thickness of a fascine is below the banquette tread. This row is held firmly in place by small pickets driven through the fascine and into the earth on which it rests.

The other rows are laid on top of the first row, and on each other, and held in place by pickets; the fascines in each row breaking joints with those in the row beneath, and are so placed as to have a slope of $\frac{3}{1}$. The projecting ends of the bands are stretched, and small pickets are also driven through the loops. These pickets are called **anchoring** pickets. Six rows of fascines would place the top of the upper row within four inches of the interior crest. A row of sods should be laid along this top row to complete the interior slope.

If five rows only are used, the portion of the interior slope between the top row and the interior crest is left to stand at the natural slope of the earth until there is time to finish it. It is usually finished by means of sods.

79. Gabions.—Gabions are basket-work cylinders, open at both ends (Fig. 25). The basket-work is made of brushwood, although iron has been used.

Fig. 25.

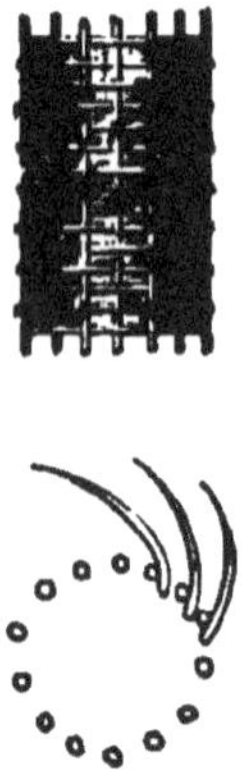

A gabion is usually made two feet and nine inches high, and two feet in diameter. The method used ordinarily in making a gabion is to drive pickets, three and a half feet long, into the ground in a circular row. The size of the circle may be determined by a circle traced on the ground, or by a circular piece of wood called a **gabion form.** The gabion form is then raised about one foot above the ground, and the pickets are bound firmly together by a rope passed around them just below the form.

The brush wood, stripped of leaves and twigs, straight and flexible, and about one-half inch in diameter, is then woven around the pickets.

The most approved method of weaving is to use three rods as shown in Fig. 25. The three rods are placed with the large ends inwards, and separated from each other by an intervening picket. The first rod (the one in rear) is passed over the other two rods, around two pickets, and within one; the second rod (now the rear one) is passed over the rods, around two pickets, and within one; the third is manipulated in the same manner. Each rod comes to the front in turn, and is separated from the others by a picket. As the weaving progresses, the basket-work must be kept pressed down, either by the foot, or by blows from a mallet or a stout stick. When the basket-work has reached the height of two feet and nine inches, the ends of the rods are firmly secured by withes. If the basket work is not of the requisite height, by reason of the pickets being driven too far into the ground, it is turned over and added to on the other end, both ends being secured by withes. The usual number of pickets is nine, but a larger number may be used when the brushwood is flexible and small.

With a form, two men can make a gabion in one hour and a half; without it, it will take three men about two hours to make one gabion.

Gabion revetments are rarely used for the interior slopes of ordinary field works. They are much and usefully employed in the trenches in siege operations, in batteries, and in embrasures.

When used as a revetment for the interior slope, they give a height of three feet, in consequence of the projecting ends of the pickets. They are made to rest upon a row of fascines half buried in the banquette, and are so placed as to have the same inclination as the interior slope. The gabions are then filled with earth, and the parapet is raised behind them.

When the parapet reaches the height of the gabion, a row of fascines is laid on top of the gabions to give the requisite height to the interior crest.

80. Hurdle revetments.—Hurdle revetments are frequently used in field works.

The hurdle is ordinarily two feet and nine inches high, and about six feet long. It differs in construction from the gabion, only in having the pickets placed in a straight line, or along a line which is slightly curved.

When used for revetting long lines of parapet, the hurdles are made continuous. The pickets are driven into the ground as close to the parapet as possible, leaving only space enough to allow the brushwood being woven in and out and around the pickets. The pickets should have the same inclination as that of the interior slope, and be tied back by anchoring pickets. A larger size of brushwood can be used for this revetment than can be used for making gabions.

81. Timber Revetments.—Fascines, gabions, and hurdles are used for revetments when timber is scarce, or when there are better uses to which the timber may be applied.

Timber is a suitable material, both on account of its comparative durability, and for the ease with which it can be worked into the shapes required. Its scarcity, and its usefulness for other purposes, are the reasons for not employing it more freely for revetments. When abundant, it will be used in preference to any of the other materials for the revetment of interior slopes and scarps.

There are two general methods employed in making timber revetments. One method is to place the logs in a horizontal position, piling them in rows one above the other with the proper inclination, and fastening them in place, by pinning the layers together, and by anchoring ties. The other method is to cut the sticks into short lengths and to place them in contact with each other in an upright position, then cap the posts by a horizontal piece of timber. (Fig. 26.)

Fig. 26.

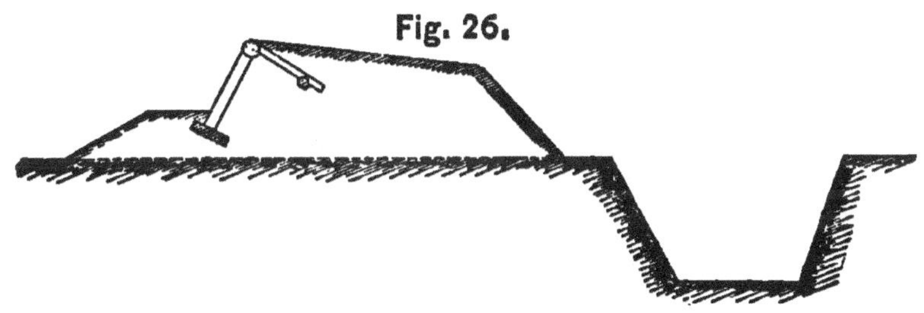

The latter method is considered to be the better one, especially in the event of a cannon shot penetrating the parapet and striking the back of the revetment. If the log struck is knocked out, the sphere of injury would be more extended when the log was in a horizontal position than if it were in an upright one.

This method of revetting the slopes (Fig. 26) was

much used in the late war in the United States. General Barnard of the United States Engineers describes it as follows:

"This (the revetment) consisted of posts from four to six inches in diameter of oak, chesnut, or cedar, cut into lengths of 5½ feet and set with a slope of $\frac{4}{1}$ in close contact in a trench, at the foot of the breast height, two feet in depth. These were sawed off sixteen inches below the crest and shaped to receive a horizontal capping piece of six-inch timber, hewed or sawed, to a half-round, as shown in the sketch."

The lower ends of the posts rested upon a two-inch plank, placed in the bottom of the trench. The capping piece was "anchored" in place, by ties notched into logs buried in the parapet. The slope of $\frac{4}{1}$ is steeper than that assumed for the ordinary profile.

82. Plank Revetments.—Plank is an excellent material for revetments where durability and great strength are not required. The ease of working and convenience of handling are its great advantages. When it can be easily obtained and can be spared for the purpose, it will always be used in works of hurried construction.

Revetments may be made with it by driving posts or pieces of scantling into the earth, three or four feet apart, giving to them the same inclination as the interior slope. Boards, in a horizontal position, to retain the earth, are then nailed to these scantlings or posts.

Or, the scantlings may be capped, and the boards having been cut into suitable lengths, placed in an upright position, similar to the posts in the timber revetment, shown in Fig. 26. The moisture of the earth soon produces rot in the boards and renders the revetment a very perishable one.

83. Casks, barrels, etc.—Casks, barrels, and materials of this description, are frequently found in numbers within the neighborhood of the work, and can be usefully employed in making temporary revetments. The construction of revetments of this kind will be similar to that in which gabions are used.

84. Sand-bags.—Sand-bags are sacks made of strong canvas, and then filled partially with sand or earth. They are used quite frequently, when the other materials for revetments can not be easily procured; or where, from peculiar circumstances, they can be usefully employed to give speedy protection against an enemy's fire. The sacks are usually made about two feet and three inches long, and one foot and two inches wide, when empty. They are about three-fourths filled with earth, and are then tied at the top.

The bags are not entirely filled with earth, so that they may be flattened when laid in the revetment. They are usually laid as "headers and stretchers," that is, the ends and sides of the bags appear alternately in each course. (Fig. 27.) The bags break joints in each course. Sometimes the bags are all laid as headers.

Fig. 27.

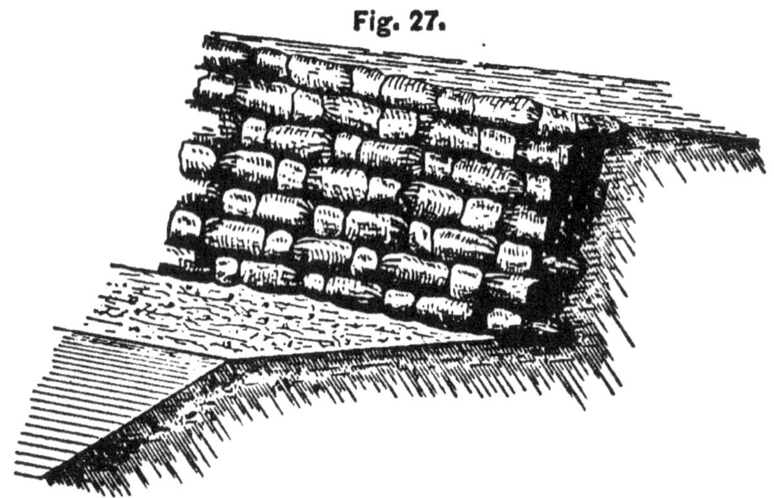

It will require eighteen sand-bags of the size just given, laid as headers and stretchers, to make one square yard of revetment.

The sand-bag revetment possesses peculiar advantages, where speedy shelter is required, or where the earth has to be carried some distance under fire. Its great defect is the perishable nature of the revetment, due to the speedy rotting of the bags.

Fig. 28.

Sand-bags are very useful to place on the parapet of a work to form improvised loop-holes for the defence. (Fig. 28.)

85. Sod revetments.—Grass sods form the best revetments for the interior slopes and for the cheeks of embrasures.

Sod revetments are durable and do not splinter when struck by the enemy's projectiles. They require considerable time in which to construct them.

Sods for revetments should be cut from fine, close turf with thickly matted roots, the grass having been mowed previous to the cutting. They are cut in the form of blocks, 16 inches long, 8 inches wide, and 4 inches thick.

The construction of a sod revetment should be commenced, when practicable, as soon as the parapet is raised to the level of the tread of the banquette. The first course is then laid with the grass side downwards, alternately as headers and stretchers; or, two stretchers to one header. The upper surface of this layer should be perpendicular to the plane of the interior slope, and this position is given to the sods by sloping the bed on which they rest.

The second course is then laid in a similar manner on the first, and breaking joints with it. Sometimes a few wooden pegs are driven through the upper sods into the lower ones, to fasten them together; but this is not necessary.

The other courses are laid in a similar manner to that described for the first and second, until the top course is reached; which should be laid with the grass side upwards.

Each course is settled firmly in place by tapping the sods as they are laid, with the flat side of a spade, and packing the earth of the parapet closely behind each layer.

It is recommended to have the sods laid protruding slightly beyond the face of the interior slope, so that each course, after it has been laid, may be nicely trimmed.

In the sod revetments used in the Washington defences, the sods were eighteen inches long, twelve inches wide, and four inches thick. They were laid to form a **sod-wall**, twelve inches thick, and with a slope of $\frac{4}{1}$. The earth of the parapet was thoroughly rammed behind the revetment, and was carried up simultaneously with the laying of the sods. The courses were connected by small wooden pegs, three-fourths of an inch in diameter and nine inches long, driven through each alternate course into the layers beneath.

86. Pisa Revetment.—A pisa revetment is a wall of clay built against a slope to protect it.

Its construction is as follows: Common earth, mixed with clay and moistened with water, is kneaded until the particles will adhere when pressed or squeezed together. Sometimes chopped straw is mixed in the mass. A row of pickets, with the proper inclination given to them, is driven along the foot of the interior slope, the tops extending a short distance above the height marked for the interior crest. A shallow trench about twelve inches wide is dug in the parapet, behind the line of pickets, and a board laid horizontally on edge on the side next to and supported by the pickets. The tempered clay is then placed in the trench and rammed. Successive layers are placed in until the clay reaches the top of the board, the earth of the parapet being carried up simultaneously with the revetment. A second board is then placed upon the first, and the clay rammed in, rising simultaneously with the parapet as in the first course; and this process is continued until the top layer is on the same level with the interior crest. When the clay has dried, the boards and pickets are removed.

87. Other revetments.—Sun-dried bricks, adobes, and stones, have been used for revetments. They are built into walls in a method similar to that described for sods.

Any material which can be used to protect the earthern slope will be suitable. Time, durability, and fitness are the things to govern in the selection of materials for revetments.

For batteries or parapets exposed to artillery fire, the fascines are preferable for revetments. Gabions are to be preferred for embrasures and traverses; sand bags for powder magazines and traverses. Sometimes all three of them are used simultaneously in the same field work.

Gabions require less wood than fascines, and are more easily made. They do not have to be tied or fastened in place, like fascines. Sand-bag revetments are more quickly made, but the canvas rots quickly. Sods are the best material, but revetments made from them require a great deal of time and labor to construct.

CHAPTER X.

DEFILADE.

88. Defilade.—The first principle, enunciated in Art. 7, imposes the condition that the defenders of a field work should be screened from the enemy's view.

If the work is built upon a level site which commands all the ground around it, this condition is fulfilled, when the space occupied by the men is enclosed by a parapet eight feet high.

If the work is built upon an irregular site, or if there are heights within cannon range which command it, a parapet of uniform height and only eight feet high will not screen the men in the interior from the enemy's view.

The arranging of a work, under these circumstances, so that the men standing on the terreplein behind the parapet shall be screened from the enemy's view, and be protected from his fire, is called **defilading** the work.

There are three ways which may be used to defilade a work.

1. By raising the parapet.
2. By lowering the terreplein.
3. By using traverses.

The labor of defilading may be reduced, and in many cases avoided, by giving proper directions to the lines of the work.

89. Plane of defilade.—To illustrate this subject, take a field work built upon a site which is practically level, but within cannon range of a hill which commands it. (Fig. 29.)

Fig. 29.

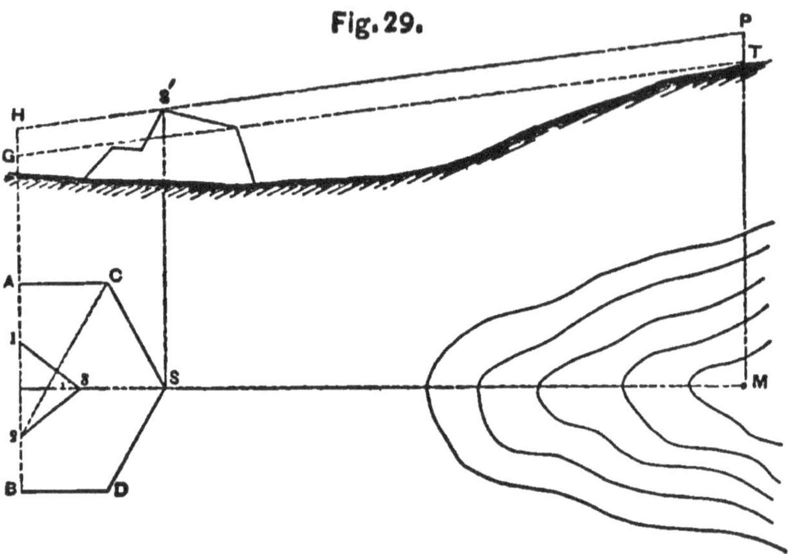

It is plain that if the parapet from C to D is not high enough, an enemy occupying a position at M can see the defenders when standing on the terreplein behind the parapets of the flanks, A C and B D.

It is also plain that if the parapet be raised high enough to conceal the men on the terreplein at A and at B, an enemy at M can not see the defenders on the terreplein in any portion of the work.

Safety for a man at A or B requires that the line of fire coming from M should pass at least six feet and six inches above the ground on which he stands (Art. 20). In other words, he must be sheltered from all fire of the enemy coming from M.

Let H P represent a line which passes eight feet above the ground along A B, and five feet above the ground at M; this will be a line of the plane of fire from M, above which the effects of direct fire from M may be neglected. Any mass interposed between the line A B and the height M, and raised until its top touches or rises above the line H P, wil. intercept the direct fire on A B, from M, and shelter the men on the terreplein at the points A and B.

If then the interior crests S C and S D are held in a plane passing eight feet above the ground at A B, and five feet above the ground at M, the parapets of these faces, S C and S D, will intercept this fire, and the work will be **defiladed.**

This plane in which the crests S C and S D are held is known as the **plane of defilade,** and may be defined to be *that plane, which containing the interior crest of a work, passes at least eight feet above those points to be sheltered, and at least five feet above the ground which can be occupied by an enemy within cannon range.*

90. Area to be defiladed.—The amount of space in rear of a parapet which is required to be defiladed, depends upon circumstances. In some cases, the entire space enclosed, and in others only a part, is to be protected from this fire from a commanding height. Thus, it is usual to require that the whole interior space of an enclosed work should be defiladed; that the interior as far as the gorge should be defiladed for a half-enclosed work; and that so much of the interior, or so much of the terreplein behind the parapet, as may be necessary for the free movements of the defence, should be defiladed in open works or lines.

91. Method of determining the plane of defilade.—It is not convenient in practice to place the eye at a distance of eight feet from the ground, nor is it an easy thing to judge, from a distance, what should be the position of a point like P, which shall be five feet above the ground. The method used is to place the eye at a convenient distance from the ground, observe the highest point of the top of the hill, and determine the position of a visual plane tangent to the hill. Knowing the position of this visual plane, a second plane is passed parallel to it and five feet above it. The tangent visual plane is known as the **rampant plane**, and the plane parallel to it is the plane of defilade.

The position of the rampant plane and the plane of defilade may be fixed as follows.

To illustrate the method, a redan is supposed to be

the field work which is to be built, upon a position commanded by a neighboring height, and that the salient and extremities of the faces are marked by upright poles planted in the ground. (Fig. 30.) The

Fig. 30.

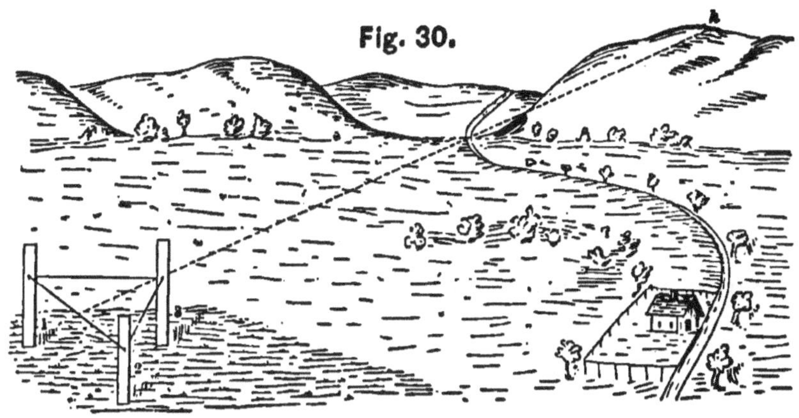

trace marked, the next step is to profile the work, and this requires the height of the interior crest to be determined.

Two stakes, at a convenient distance apart, are erected upon the gorge line; if not too far apart, the poles already erected to mark the extremities of the faces may be used. A line, three feet from the ground, is marked by a strip of wood having a straight edge, or by a cord tightly stretched, and fastened to these uprights.

An observer is placed in rear of this line; he sights along it and tangent to the hill, *h*, and determines where the visual plane containing this line cuts the pole placed at the salient. This point is marked, and with the line joining the two uprights on the gorge line, fixes the position of a plane tangent to the hill and three feet above the ground at the gorge.

If on the three posts, 1, 2, 3, points be marked, five feet above the points of intersection of the posts by the rampant plane, these will be points of a plane which will pass eight feet above the ground at the gorge and five above the ground at *h*. If the faces of the redan are held in this plane, the whole interior of the redan will be defiladed from this hill, and the last plane determined will be the plane of defilade.

The extremities of the faces at the gorge have parapets of the ordinary height, viz., eight feet; the parapets from these points, increase in height until the salient is reached, where the height is the greatest. The height of the interior crest can then be determined, at the points where the profiles are to be placed.

In the previous chapter, the height of the parapet is supposed to be uniform throughout. The site being level, there is no reason why any one part of the interior crest should be higher than another. It is nevertheless the practice, even in this case, to give additional height to the parapet at a salient, not for the purposes of defilading the interior, as just explained, but to lessen the effect of any enfilading fire which an enemy might obtain upon the faces, and to allow for the descent of the trajectory of a projectile which might graze the interior crest at the salient.

92. A slight deviation from the method just described is made when the work to be defiladed is a lunette, instead of a redan (Fig. 29).

Two uprights, about twelve feet apart, are planted upon, and near the centre of, the gorge line. A third upright is placed in front of the gorge and ten or twelve feet from it, upon the line joining the centre of the gorge line with the salient.

The points are then marked where the rampant plane, three feet above the ground, cuts these three uprights and the uprights planted at the salient, **S**, and at the shoulders, **C** and **D**. A distance of five feet is marked above the points just determined, and this will fix the position of the plane of defilade for the lunette.

93. Front and reverse defilade.—The supposition has been made that there is only one commanding height within cannon range of the field work. Suppose that there are two or more heights, from each of which an enemy could see the interior of the work, if an ordinary uniform height of parapet is used.

Take the case of a redan. There are two distinct conditions which may arise. One of them is where all the heights will lie within the angle of the faces prolonged; the other is when one or more of these heights will lie without this angle. Thus, the redan,

A S B, (Fig. 31) has the heights Z and Z′, lying within the angle of the faces prolonged; and the heights, X and Y, lying without this angle.

Fig. 31.

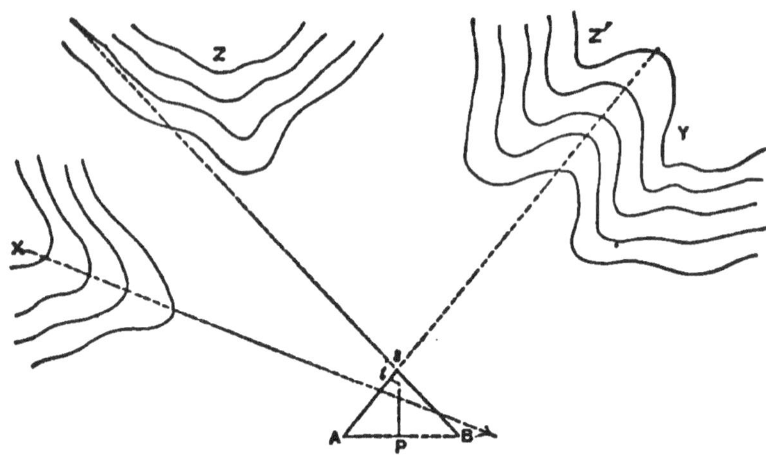

The heights Z and Z′ can only bring front fires, direct or slant, upon the faces of the redan. The height X can bring a direct fire upon the face A S, and a reverse fire upon the face B S. A similar condition exists for the height Y.

The arrangements made to screen the terrepleins from the front fire are all made like those just described, and give rise to the problem known as **front** or **direct defilade.** Those made to screen the terreplein and the men on the banquettes from reverse fires give rise to the problem of **reverse defilade.**

The method used to defilade a work from two or more heights, when these heights can only bring front fires, is the same in principle as that used for a single height. It is only necessary in a case of this kind to defilade the work from that height which gives the highest fire.

When some of the heights expose the work to a reverse fire, it is usual to defilade the faces separately from the front fire, and then make the necessary arrangements to protect the interior against reverse fire.

94. Method used to defilade a work exposed to reverse fire.—The first step, as has just been stated, is to defilade the work from the front fire. It is evident, even when this is done, that men upon the banquette are exposed to reverse fire.

Thus, suppose a redan like that in Fig. 31, to have had its faces defiladed from front fire. It is plain that men, standing on the banquette at B, would be above the plane of defilade containing the face A S, and would be seen by an enemy at X. Suppose the interior crest of the face, A S, to be raised high enough to shelter the men on the banquette of the opposite face,

Fig. 32.

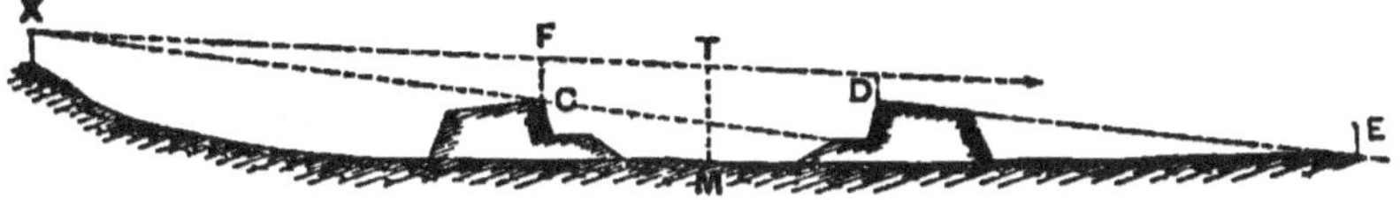

B S. The effect would be to shelter the men there, but to expose the men on the banquette of the face A S. Thus, if the crest at C, (Fig. 32), is raised to shelter the men on the banquette at D, it exposes the crest at F to a fire coming from E. The defect of reverse fire is not removed, but only transferred to another place by raising the parapet. This method can not be used.

The method that is used is to interpose an auxiliary mass which shall intercept the dangerous reverse fire. A mass of earth, placed across a line of fire to intercept the missiles moving along that line, is termed a **traverse**. Thus, if a traverse of earth is placed between the banquettes and made thick enough to protect the men, and high enough to intercept the enemy's missiles, the men can stand on the banquettes without fear of the reverse fire.

The thickness of a traverse, like that of the parapet, depends upon the penetration of the enemy's projectiles.

The height of the traverse depends upon the height of that plane of reverse fire, above which the fire may be neglected. This plane which determines the height of the traverse is known as the plane of **reverse defilade**. It may be defined to be, *that plane which passes at least two feet above the interior crest exposed to reverse fire, and at least five feet above the ground from which this fire can come.*

When both faces of the work are exposed to reverse fire, there will be two planes of reverse defilade. In this case, the least height of the traverse will be determined by the intersection of these planes of defilade, and the line of intersection will form the middle line of the top of the traverse. If it be more convenient to place the traverse along some other line, the capital for instance, the height of the traverse will be determined by the higher of the two planes.

A traverse placed along the capital frequently crowds the salient too much; in which case, its direction is changed when near the salient, and it is joined to the face most exposed. (Fig. 31.)

95. Method of determining the plane of reverse defilade.—Since a traverse is to be used, it will be supposed that its place is along the capital. This traverse will divide the interior space into two equal parts. It will only be necessary to defilade each half from the front fire; which may be done as follows:

Uprights are placed along the capital to mark the middle of the traverse. A line, three feet from the ground, is marked upon these uprights by stretching a cord, or by nailing strips of wood to the poles. The rampant planes passing through this line, and tangent to the hills on each side which furnish the highest front fire, are then determined, and their intersections with the uprights placed along the sub-crests of the faces, are marked. On these verticals the planes of front defilade are then marked, and these points, thus obtained, determine the height of the interior crests of the faces. The profiles are then placed along the sub-crest at the proper points.

The next step is to determine the position of the planes of reverse defilade.

To illustrate the method, it will be shown how the plane of reverse defilade is found for the face, S B, of the redan in Fig. 31, which is exposed to a reverse fire from the high ground at X.

It is supposed that the faces, S A and S B, have been defiladed from front fire by the method just described; that the profiles of the parapet have been erected; and that the vertical poles, marking the position of the middle line of the traverse, are in place.

Two, or more of the profiles along the face S B are taken, and are joined by a straight line which is just three feet below the interior crest of this face, measured on these profiles. The position of this line is fixed usually by marking, on the vertical on the sub-crest of each profile, a distance of eighteen inches above the banquette, and by joining the points thus determined by a strip of wood or a tightly stretched cord fastened to the profiles.

An observer is placed behind this line and determines the position of a visual plane which contains the line, and is tangent to the hill, X. The points of intersection of this visual plane with the poles placed to indicate the position of the traverse, are marked.

These points marked on the poles, with the line three feet below the crest, fix the position of the rampant plane for the reverse defilade. The distance of five feet above this plane, is marked upon these poles, and the points marked thus will be points of the plane of reverse defilade, and will determine the height of a traverse required to protect the face B S from the reverse fire from X.

A similar process determines the plane of reverse defilade for the face A S.

The position of the higher plane will determine the height which is to be given to the traverse. The form of a traverse and its construction will be mentioned hereafter.

96. Defilade by traverses.—The problem of reverse defilade is almost always solved by the use of traverses. That of front defilade, by either raising the parapet or lowering the terreplein. The latter plan is the method pursued in permanent fortifications; it is sometimes used in ordinary field works, but rarely, because the terreplein and surface of the ground are generally one and the same.

It is not always practicable to solve problems of front defilade in field works by raising the parapet, because the height required for the parapet to reach the plane of defilade may be too great, and exceed the limit laid down. In these cases, traverses are used.

As an example, take a lunette commanded by a

hill, as X (Fig. 33) and suppose that, with a command of twelve feet at the salient, the interior crest protects only half of the interior of the work.

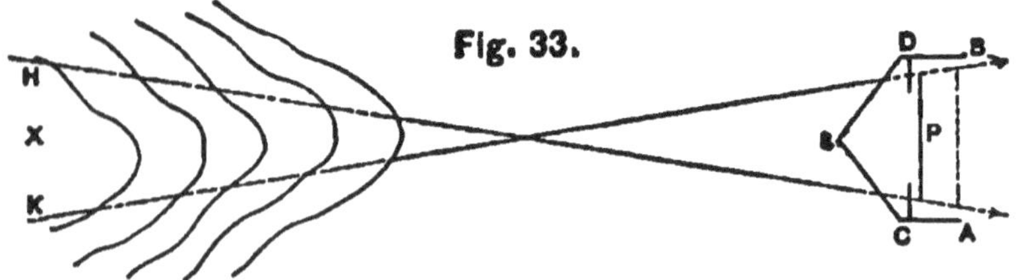

Fig. 33.

The remaining half can only be protected by means of a traverse, P, placed along the line dividing the part which is defiladed from that which is not. Two short traverses would be placed near the shoulder angles, overlapping the ends of the principal traverse, as shown in the figure, with room between the traverses to allow of a passage-way.

The short traverses at C and D would be used, even if P were not, to protect the men on the banquettes of the flanks from slant reverse fires coming from H and K.

If this traverse P is placed in the work to shelter the men from a fire in the opposite direction, which would be a reverse fire upon the faces, it is termed a **parados**. a name employed to designate the traverse when it is used to intercept a reverse fire of this kind.

The great objections to the use of traverses are the space they take, and the labor they require for their construction. When built they are often utilized in other ways, besides the primary use as an intercepting mass.

97. Defilade of lines.—The same methods may be used to defilade an open work, a part of a line, or a line, as are used to defilade a half-closed or a closed work.

The longest sides, or the faces, of an open work or a line should be traced, where practicable, perpendicular to the direction of the enemy's fire; and should be located, if possible, so that the parapet shall occupy a ridge, or be so situated that the ground in rear of the parapet shall slope away from it.

A line traced and placed in this way will, for the same height of parapet, defilade a larger extent of terreplein than it would if its direction were inclined to that of the fire, or the parapet were placed upon ground which was level, or which sloped towards the enemy.

When the commanding ground is a ridge or suc-

Fig. 34.

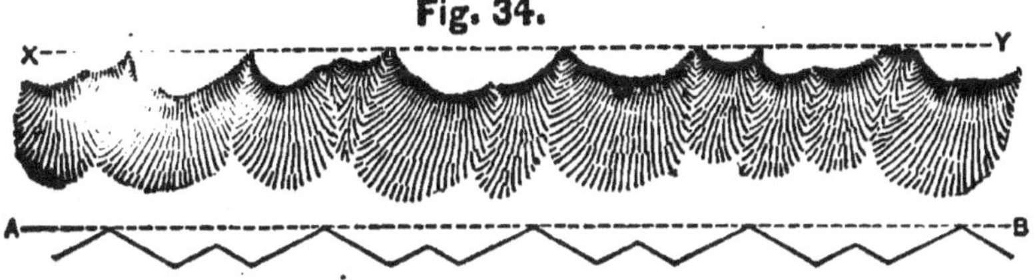

cessions of hills on the same general level, the direction of the line should be parallel to the general direction of the ridge. (Fig. 34).

The simple right line with a uniform relief will, in this case, defilade the terreplein more efficiently than if the line were broken, since the general direction of the line is perpendicular to the general fire coming from the heights.

If, however, a broken line is to be used, a continued redan line for example, it is recommended to make the salients of the large redans as obtuse as possible. By so doing, the perpendicular distance between the salients and re-entrants is shortened, and the long faces are made more nearly perpendicular to the enemy's close fire.

If the ridge is not too high, and the distance from the salient to the re-entering is not too great, it will be practicable to hold the interior crest of the line in a plane of defilade, and not have the height of the parapet too great at the salients. Moreover, the obtuseness of the salients will cause the prolongations of the faces to intersect the high ground at distances so great, that the enfilade fire from these points may be disregarded. If the heights are not on the same general level but slope downwards to the plane of site, a better position for the line is one which is oblique to the crest of the ridge. A direction should be given to the line, so that its *front*, prolonged, would pass through the point in which the meridian line, or vertical contour, of the ridge pierces the plane of site. Thus, a ridge of high ground, sloping downwards from X to Y (Fig. 35), commands the ground, A B, on which a line of fortification is to be built.

Fig. 35.

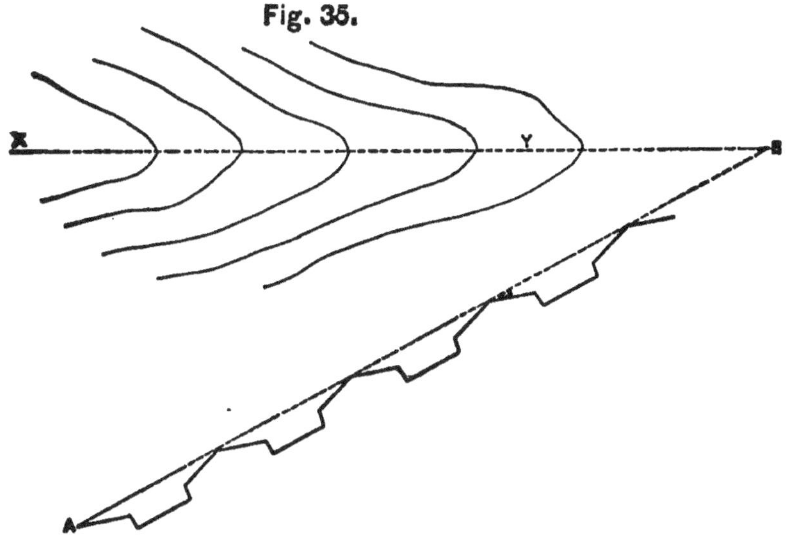

The vertical contour line of the ridge is the intersection of a vertical meridian plane with the ridge, which line is projected upon the plane of site in the dotted line X Y, and pierces this plane, in this case, at B. The front of the line should have a direction like A B.

In this position a plane can be passed in which the salients, with a uniform command, may be held, and the terrepleins be defiladed from the high ground in front.

A straight line of parapet would, in this case, afford the best cover for the terreplein behind the parapet, as it would in the case where the line is taken parallel to the commanding ground.

But, if a broken line be used, care should be taken, as in the last case, to make the salients obtuse angles. If the line be a bastioned line, it is plain that the flanks are liable to be exposed to both reverse and enfilading fires.

The nearer the angle A B X is to a right angle, the better will be the position of the line A B to allow of its terreplein being defiladed from the high ground in front.

CHAPTER XI.

THE INTERIOR ARRANGEMENTS OF A FIELD WORK.

98. Classes.—The earth work for the parapet being completed, and the revetments of the interior slope constructed, attention is then paid to the interior of the work. Certain arrangements have to be made in the interior, to add to the efficiency of the defence, and to provide for the comfort of the troops who have to occupy the work. These interior arrangements are divided into classes, according to the object to be attained by them.

The divisions may be classified as follows:

1. The arrangements of, and along a parapet, intended to add to the efficiency of the defence;

2. The arrangements within the area enclosed by a parapet, to shelter the men and *matèriel* from the fire of the enemy:

3. The arangements made to allow egress and ingress of the troops; including those made to guard the outlets against surprise; and

4. The arrangements which may be made to provide for the comfort and welfare of the garrison when occupying the interior of the work for some time.

I. Arrangement of the Parapet.

99. Defence.—The work may be defended by musketry alone, or it may be defended by artillery combined with musketry.

The arrangements of the parapet for musketry are completed when the banquette and the revetment of the interior slope are finished.

The work, in this condition, does not admit of the use of artillery. Some additional arrangements must be provided, if artillery is to be employed. The fire of artillery is either over the parapet or through it. In the former case, the fire is said to be "*en barbette;*" in the latter, it is called "*embrasure fire.*" In both of these fires, arrangements must be made in or along the parapet for the service of the guns.

100. Barbette fire.—Barbette fire can only be obtained by some arrangement which raises the gun into a position from which it can be fired over the parapet. There are two methods by which this can be done; *one,* by mounting the gun on a high carriage, or on a carriage which admits of the gun being raised to the necessary position; *the other*, by building a mound of earth sufficiently high behind the parapet, and placing the gun on this mound. The latter is the method generally employed in field works.

This mound of earth in rear of the parapet, with its upper surface arranged so that a piece of artillery placed upon it can fire over the parapet, is called **a barbette**.

The artillery used, in the defence of field works, may be either siege or field guns; but most generally the latter are employed.

The upper surface of the platform on which the wheels rest, or the upper surface of the barbette, when no platform is used, should be at a distance below the interior crest just sufficient to allow the gun to fire over the interior crest and parallel to the superior slope. A distance greater than this would interfere with the efficiency of the gun; a distance less, would unnecessarily expose the gun and carriage to the enemy's fire.

The axis of the trunnions of a field gun, in the United States service, is about forty-three (43) inches above the ground on which the wheels rest. The diameter of the piece at the muzzle and the inclination of the superior slope being given, it is easy to determine what should be the distance of the upper surface of the barbette below the interior crest.

The general rule used is to take this distance at two feet and nine inches for field guns, and four feet for siege guns, when mounted upon the ordinary carriages.

101. Construction of the trace of a barbette.—The trace of a barbette for a single field gun behind the parapet of one of the faces of a work, may be constructed as follows:

Let A B (Fig. 36) be the interior crest of the face, and C the point at which the barbette is to be constructed.

Fig. 36.

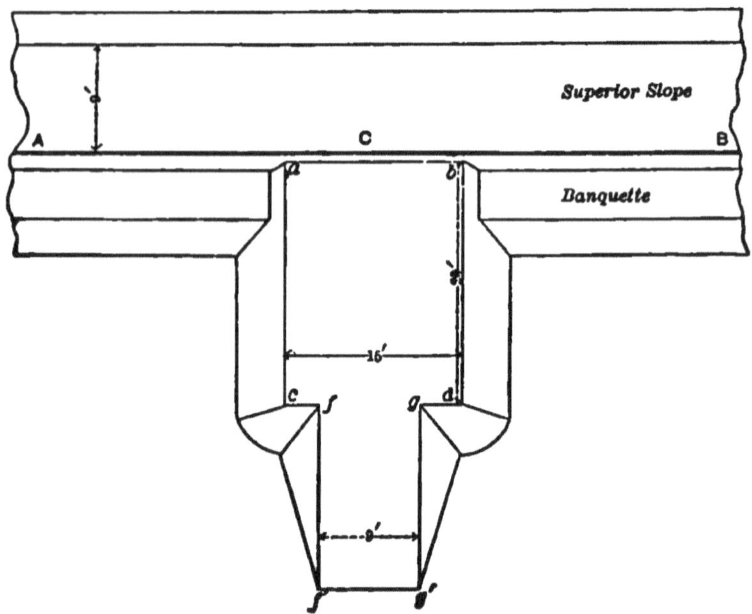

The upper surface of the barbette is to be made level, and large enough to allow the working of the gun.

A field gun firing over a straight parapet requires a front of ten feet, at least, and should have a depth of twenty feet, to allow for the recoil and proper handling of the piece. Since the upper surface of the barbette is thirty-three inches below the interior crest, its intersection with the interior slope will be a straight line, *a b*, parallel to A B, and eleven inches from it in projection. Make *a b* equal to the required length of front (taken to be fifteen feet in this case) and through the extremities, draw the lines *a c* and *b d* perpendicular to *a b*, making them each equal to twenty feet. The area *a b c d* will be the upper surface of the barbette, in plan. The upper surface of the barbette is joined to the site by the natural slope of the earth, which is assumed to be $\frac{1}{1}$.

The height of the parapet is eight feet, which fixes the height of the barbette at five feet and three inches. The intersections of the natural slopes of the sides and rear with the plane of site are then easily determined, as well as the intersections with the banquette tread and the banquette slope. These intersections determined, the construction of the trace is completed.

The gun is carried to the top of the barbette by an inclined road, called a **ramp**, connecting the upper surface of the barbette with the plane of site. The ramp, *f g f' g'*, has a width of nine feet and a slope of $\frac{1}{6}$, and is here placed at the rear of the barbette. The width, the slope, and the position, are all governed by convenience and the circumstances of each case.

The sides of the ramp are joined to the site by the natural slope of the earth.

102. Barbettes, when used, are almost always placed in the salients. Their construction is practically the same as when placed against the face of a work. It is usual, when a barbette is made in a salient, to fill the salient angle for a short space with earth, and to form a short face, called a **pan-coupé**, (Fig. 37),

Fig. 37.

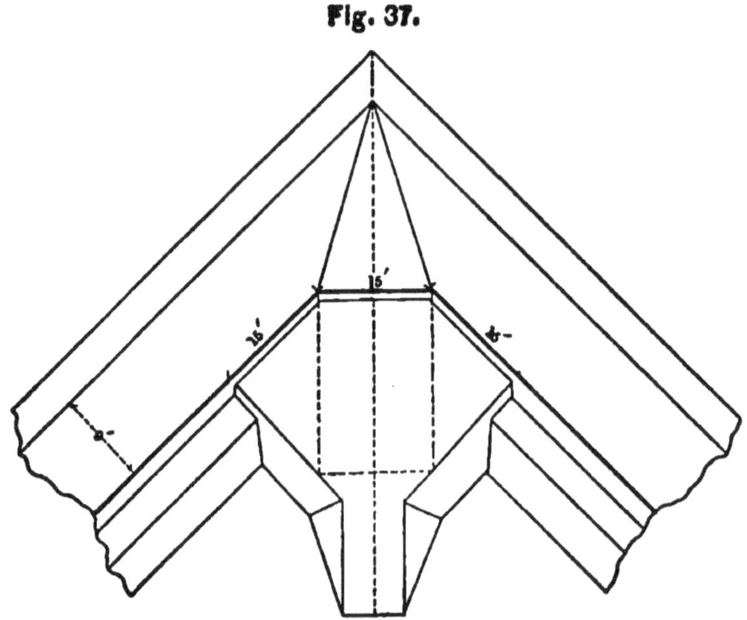

over which the gun fires in the direction of the capital. The length of this pan-coupé is from ten to fifteen feet.

The construction of the plan differs from the one described only in the form of the upper surface. In this case, the upper surface is pentagonal in form, care being taken to make it large enough to allow the gun to be fired over the faces of the salient, as well as along the capital.

In this particular case, it is arranged for only one gun. It may be arranged for several.

103. Embrasure fire.—The opening made in a parapet to allow a piece of artillery to fire through it, is called an **embrasure.**

The gun may rest upon the natural surface of the ground in rear of the parapet, or it may be placed upon a mound of earth similar in shape to a barbette, only not so high.

The bottom of the embrasure, in either case, must be placed so that the gun, resting on its platform can fire through the embrasure when run "in battery." The bottom of an embrasure is called **the sole.** It is given an inclination, to allow the gun to be fired with its muzzle depressed. This inclination is usually taken to be the same as that of the superior slope of the parapet, unless otherwise stated.

The opening, *a b* G H, (Fig. 38), in the interior slope is called the **throat,** and should be made as small as practicable. It is usually made from eighteen to twenty-four inches wide, and either rectangular or trapezoidal in form.

The embrasure widens towards the exterior, the amount of widening depending upon the extent of field of fire required. This widening is called the **splay.**

The sides of the embrasure are called the **cheeks**; the line, M N, which bisects the sole is called the **directrix.**

The portion of the interior slope which is below the throat of the embrasure is termed the **genouill-ere**; the mass of parapet between two embrasures is termed a **merlon.**

Fig. 38.

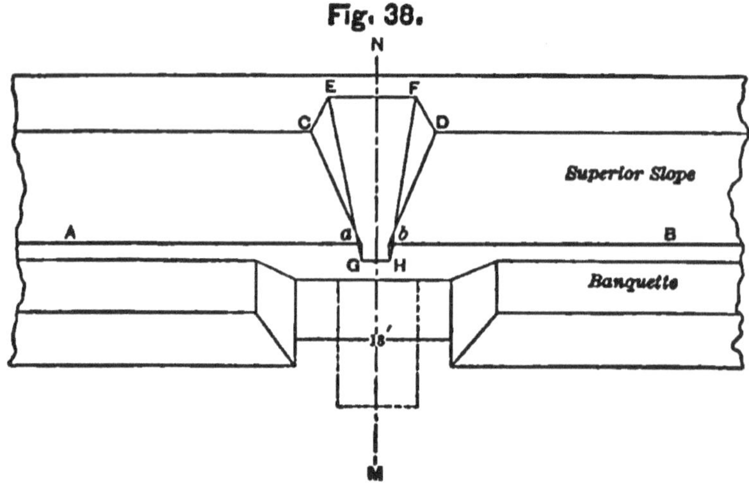

The embrasures in field works are usually cut after the parapet has been constructed, and, in important works, the exterior openings are **masked** until the moment to use them arrives, to prevent their position being discovered by reconnoitering parties of the enemy. A detail of six men should be able to cut an embrasure in the parapet of a field work and finish it in eight hours.

104. Trace of an embrasure.—To make the trace of an embrasure, draw the line M N. (Fig. 38), perpendicular to the interior crest, at the point where the line of fire of the gun intersects it, and set off *a b* equal to the width of the throat at the top. If the throat is to be rectangular, draw through *a* and *b* the lines *a* G and *b* H, perpendicular to the interior crest, and produce them until they intersect the line, G H, drawn parallel to the interior crest and at a distance from it equal to one-third of the depth of the throat. This gives the plan of the throat.

To trace the plan of the sole, the inclination and splay must be known. Knowing these, the intersection of the sole with the exterior slope can be determined. This intersection may be determined as follows:

Construct a profile of the parapet, and find the point in which the line G H pierces it. Through this point thus determined draw a line in the plane of the profile, making the same angle with the horizontal that the sole makes with the horizontal plane, and find the point in which this line intersects the exterior slope. This point of intersection is a point of the line E F, which line is parallel to the exterior crest.

The splay of the sole is usually determined, in plan, by giving to E F some definite length, and then joining its extremities with the lower line of the throat. A throat twenty inches wide will have a horizontal field of fire of twenty-two degrees, when E F is equal to one half the thickness of the parapet; a fire of thirty-one degrees, when the E F is equal to two-thirds of the thickness; a fire of forty-eight degrees, when this line is equal to the thickness of the parapet.

Supposing twenty degrees to be the field of fire required, E F is made equal to half the thickness of the parapet, and the points E and F are then joined to G and H, which completes the plan of the sole.

The cheeks are determined by setting off on the exterior crest, on the right and the left of the sole, points as C and D, the horizontal distances of which from the lines E G and F H, measured on the line C D, shall be equal to one-third of their height above the sole; and then joining these points to *a* and *b*, and to E and F, by straight lines. These lines complete the plan of the embrasure.

The cheeks are warped surfaces which may be generated by moving a straight line, touching the sides of the throat, and the sides of the opening in the exterior slope, so that the intersection of the surfaces with the sole and the superior slope shall be straight lines.

The cheeks should be revetted for several feet at least from the throat, to protect them from the weather and the blast of the guns. Gabions are usually employed for this purpose, and are covered in action with raw-hides when they can be procured. Sods make good revetments for the cheeks, if there is time to finish them.

Consecutive embrasures should not be nearer to

each other than fifteen feet from centre to centre, to prevent crowding of the guns, and to prevent the merlon, **M**, (Fig. 39) from being too weak. A merlon

Fig. 39.

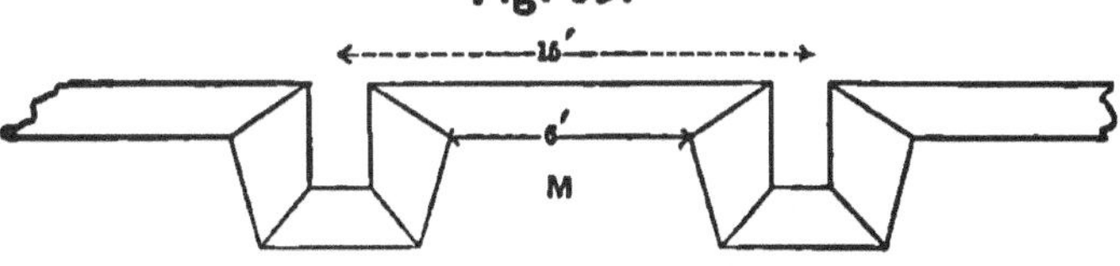

which measures less than six feet on the exterior crest should not be allowed, as it would make the parapet too weak.

105. Direct and oblique embrasures.—A **direct** embrasure is one in which the directrix is perpendicular to the interior crest at the point of intersection with the crest; an **oblique** embrasure is one in which the directrix makes an angle with the interior crest. When possible, direct embrasures are the ones which are made. If oblique embrasures are to be made, their method of construction is practically the same as that given for direct embrasures.

Oblique embrasures do not admit of the muzzle of the gun being inserted so far as the direct ones, and they weaken the parapets more.

Oblique embrasures are not used, as a rule, if the directrix makes with the normal to the crest an angle exceeding ten degrees. In case the angle is greater, the embrasure is provided for, in field works, by modifying the interior crest by means of the method known as "**indenting.**"

This method consists in making the crest a crémaillère line, instead of a right line, with the short branches perpendicular to the direction of the fire, and in these short branches constructing direct embrasures.

106. Comparative advantages and defects of the two kinds of fire.—The advantages of barbette fire in a field work consist in its wide field of fire, its commanding position, and its not weakening the parapet to obtain this fire. The main defect is the exposure of the men and guns to the enemy's fire.

The advantages of embrasure fire over the barbette are, that the men serving the guns, and the guns themselves, are both protected to a greater degree from the enemy's fire. The defects are, its limited field of fire, its weakening the parapet to obtain this fire, the constant repairing of the cheeks of the embrasures when in use, and the good mark the embrasures offer to the enemy.

107. Bonnettes.—It is frequently desirable that the height of the parapet, at certain points, should be increased for a short distance. This increase is generally obtained by making use of the constructions known as **bonnettes**. A bonnette extends but a short distance along the parapet, is made of earth, and is used generally to give greater protection to the men standing on the banquette against a slant or an enfilading fire of the enemy.

Bonnettes are placed usually on the salients; they are sometimes placed on the parapet between guns "*en barbette.*"

They may be constructed during the progress of the work, or after the work has been finished. In the former case, their construction is, to all intents and purposes, similar to that of the parapet. In the latter case, they are constructed generally in haste, and sand bags or gabions filled with earth are used to build them.

108. Loop-holes.—Troops on the banquette, when in the act of firing their pieces, are frequently exposed to the fire of the enemy's sharp-shooters. Under these circumstances, expedients must be devised to protect the men, without interfering with their fire. The expedient which is most generally used, is that of an improvised **loop-hole.** The loop-hole is made, in this case, by arranging two or more rows of sand bags, placed upon the parapet and filled with earth, so that the top row will be higher than the men's heads, and so as to leave intervals between the bags in the lower rows, through which the men can aim and fire their pieces. (Fig. 28.) Gabions are also used for a similar purpose. The gabions are placed in pairs upon the parapet and filled with earth, each pair being separated from the adjacent pair by an interval of about two inches.

A contrivance adopted in the war of 1861–5, was

quite effective for the same purpose. Skids were placed upon the parapet, with notches cut in them. A heavy log was placed on the skids, occupying a position parallel to the interior crest and just in contact with the superior slope. Notches were cut in the under side of this horizontal log and these were used as loop-holes. The openings to the exterior were made as small as possible, and in some cases were protected by small plates of boiler iron spiked upon the log. When exposed to artillery fire, earth was banked against the log.

A wooden loop-hole was devised by Lieut. King (now Major) of the United States Engineers, which was used in 1864. It was practically a wooden hopper made of boards, placed upon the superior slope of the parapet, and covered with earth. The splay of the sole and the angle of the cheeks were made to suit the field of fire required.

The exterior orifice of a loop-hole for musketry should be made as small as possible. A width of *two* inches and a height of *five*, is sufficiently large for ordinary purposes. The sides are sloped, and an inclination given to the bottom and top, according to the field of fire which is to be swept.

Embrasures are sometimes protected in a manner similar to this arrangement for loop-holes. Timbers are laid across the embrasure, covering the throat, leaving only room for the muzzle of the piece. These timbers are then covered by sand bags, by fascines, etc., to make them shot-proof. Sometimes the embrasure is filled in with sand bags or fascines to mask it, these things being quickly removed when the embrasure is needed for use.

Thick wooden shutters, made bullet-proof, and placed on vertical axes, and iron shutters swung on horizontal axes, have both been used to close the throat of the embrasure.

In some cases, timber supports were extended back from the parapet and a covering of timber and earth placed upon them, protecting the gun from vertical and plunging fire. A gun thus sheltered is said to be **case-mated.**

109. Traverses.—The traverses constructed along a parapet are of two kinds, viz., the traverses built to afford shelter against slant and enfilading fires, and those built as a protection against the fragments of bursting shells.

Traverses may be built at the same time that the work is constructed, or they may not be built until there is an immediate necessity for them.

In the former case, their construction is in all things similar to that of the parapet, viz., tracing, profiling and execution. In the latter case, they are generally built in great haste, and profiles are not used. The construction is of the simplest kind, having for its object to interpose a mass of earth upon a line of fire, in the shortest time possible. This is done by piling sand-bags, filled with earth upon the spot to be occupied by the traverse, and raising there a mass thick enough and high enough to serve the end required. Gabions filled with earth are frequently used for the same purpose.

The top of a traverse is usually made ridge-shaped, so as to carry away the rain water which falls upon it. The sides of the traverse are sloped, the inclination of the slopes being the same, or different, according to the degree of exposure of the traverse to the enemy's fire.

The traverse shown in Fig. 40 is an example of a

Fig. 40.

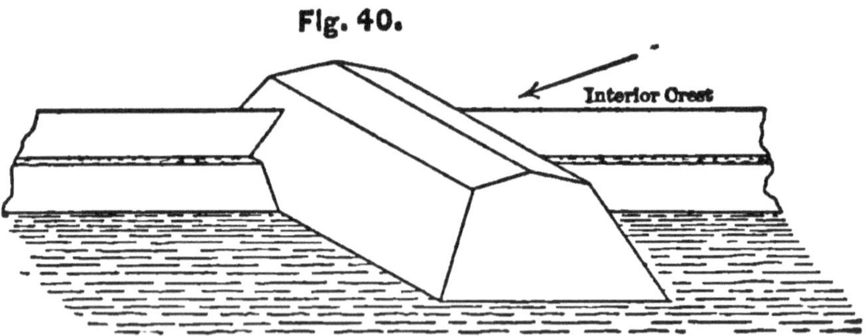

traverse built to shelter the men on the banquette from a slant or enfilading fire, coming in the direction shown by the arrow. Its top is made ridge-shaped. The side towards the enemy has the natural slope of the earth; the opposite side is made steeper, and should be revetted. The thickness of the traverse depends upon its exposure to the enemy's fire. If a fire can be brought directly upon it, it should have the same thickness as that given to the parapet.

Its height and length depend upon the amount of banquette and terreplein which are to be defiladed by it.

The manner in which this traverse is joined to the parapet is shown in Fig. 41, which represents its plan.

Fig. 41.

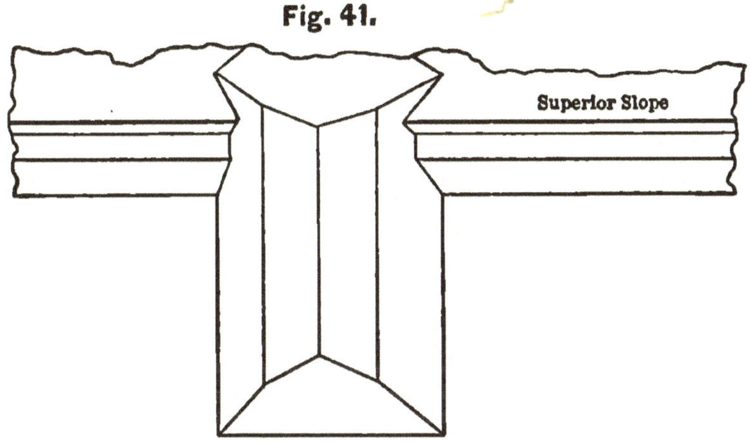

The slope on the side toward the enemy is shown, in both of these figures, to be uniform. This is not always the case. The portion exposed to the enemy's fire is given the natural slope of the earth; but below this plane of fire, the slope may be revetted, and made steeper.

Instead of being ridge-shaped, the traverses are, in many cases, made with a cross section similar to that of the parapet.

110. Splinter-proof traverses.—A traverse

intended to be used only as a protection against splinters and the fragments of shells scattered around by their explosion, is known as a **splinter-proof traverse.**

Traverses of this kind are not made so thick, nor so high, as the traverses just described. Their usual height is the same as that of the parapet. Their thickness at the base is from seven to eight feet. Their length varies, being in some cases only ten feet, and in others as much as sixteen feet.

As a rule, a traverse of this kind is not joined to the parapet, but is separated from it by a narrow passage which can be used by the men to pass from one side of the traverse to the other.

Fig. 42.

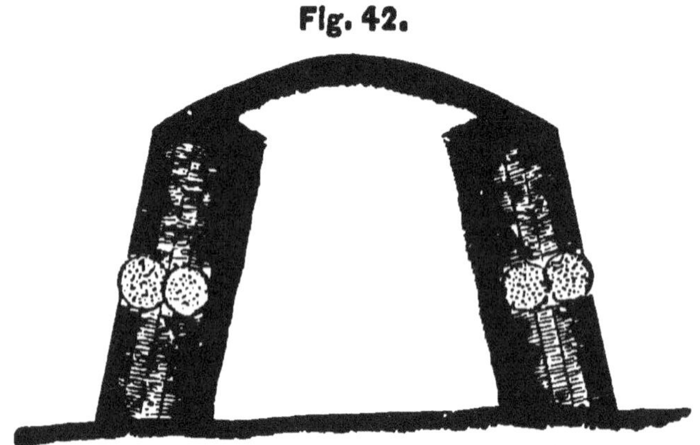

A splinter-proof traverse (Fig. 42) is usually constructed as follows:

A rectangular space is marked upon the ground for the base of the traverse. A row of gabions is then placed in juxtaposition along the line representing the base of the traverse, and given a slope inwards, either by setting the gabions on a slightly inclined excavation in the ground, or by raising the outer edges by means of fascines laid along on the ground.

The gabions are then filled with earth, and also the interior space enclosed by them.

When the earth has risen above the top of the gabions, two rows of fascines are laid upon the top of the gabions to form a base for a second row of gabions. This second row is then filled with earth, and the process of filling with earth goes on, until the earth rises high enough. The top is rounded off, or made ridge-shaped, and the traverse is completed.

The same method may be used for the construction of traverses required for defilade, when there is a pressing emergency for them.

Splinter-proof traverses are placed between the guns along a line of parapet which is exposed only to a direct fire from the enemy, and are only intended to confine the effects of bursting projectiles to a limited space.

They are usually constructed only when there is a necessity for them, and then hastily. Gabions, sand bags, fascines, or any of the materials used for revetments, may be employed in their construction.

111. Platforms.—A field gun or any piece of artillery, after repeated discharges in the same direction, soon wears the ground under its wheels into ruts, if these wheels rest upon the ground. The result of this is to lower the piece and to increase the difficulties of handling the gun.

By resting the wheels upon a hard and smooth surface this trouble is avoided. This surface is furnished, in field works, by **wooden platforms**, upon which the wheels of the gun-carriages rest. The platform is made rectangular or trapezoidal in plan, and large enough for the service of the gun.

A field gun should have a platform at least ten feet wide in front, and fifteen feet long. A siege gun would require a platform fourteen feet wide and eighteen feet long.

The dimensions, bill of materials, and method of constructing platforms, are laid down in the manuals of engineering and ordnance, and to these works the student is referred for the details of such construction.

A temporary wooden platform, which can be quickly constructed and can be used until a better one is provided, may be made as follows :

The earth upon which the platform is to rest should be thoroughly rammed. Trenches, as many as there are **sleepers**, are then dug parallel to the directrix of the embrasure, or perpendicular to the interior crest in case of a barbette. In these trenches sleepers are bedded, so as to have their rear ends raised about six inches above the front ends. Planks, at least three inches thick, are nailed or spiked transversely upon these sleepers. This practically completes the platform.

A piece of timber, at least six inches in diameter, is spiked to the front part of the platform in such a position that, when the wheels of the carriage rest against it, they will not touch the interior slope of the parapet. This piece against which the wheels rest is called a **hurter**.

It is frequently advisable to spike timbers along the sides, to act as guards and prevent the wheels of the carriage from running off the platform. These pieces, termed side-rails, are also useful to assist in holding the planks in place.

The raising of the sleepers at the rear ends gives to the platform an inclination to the front, which assists in running the gun "in battery," after it has been loaded, and assists in checking the recoil of the piece, when it is fired.

The number of sleepers used for the platform depends upon the size of the platform and the abundance of timber in the vicinity of the work.

A sleeper is laid directly under the middle line of the platform. The others are laid parallel to this one, on either side, and at equal distances apart. In emergencies, and where timber is scarce, only three sleepers may be used; a middle one, and two outside ones, the latter being placed so as to be directly under the wheels of the carriage, when the gun is "in battery." If timber is plentiful, they are placed not more than two feet apart, from centre to centre, and sometimes even in juxtaposition. At the outer end of each sleeper a stout wooden picket is driven to keep the platform steady.

In the Franco-Prussian war of 1870–1, the Prussians made use of a device which was sent into the field with each siege gun. This device consisted of two inclined planes of stout plank, faced with sheet iron, with a rise of one on six, and from eight to nine feet long. An ordinary wooden platform, or merely a platform of planks, was laid, and on it and under the wheels were placed these inclined planes.

II. Shelters for the troops, etc.

112. Shelters.—An efficient defence of a field work is greatly aided by shelters, arranged for the men and the stores, so that the men can rest in them, and the stores can be kept safe from the enemy's fire.

The shelters generally used are known as **bomb-proofs**, and **splinter-proofs**, which differ from each other only in capacity and strength.

Bomb-proofs must be strong enough to resist the effects both of the impact and of the explosion of the projectiles which strike them. They should be roomy, and when used by the men, should be well ventilated.

Splinter-proofs are so placed that they are not exposed to the impact of projectiles. They are liable to be struck by fragments of shells, or splinters knocked off by the impact of a projectile, and are therefore made only strong enough to resist the effects of the flying fragments and splinters produced by shells bursting, or by projectiles striking near them.

113. Construction of bomb-proofs.—Bomb-proofs may be built during the construction of the parapets, or after the parapets are finished. The latter is the more usual method.

The position in a field work occupied by a bomb-proof depends upon the size of the work, the kind of trace, degree of exposure of the interior of the work, the convenience of the position, etc. Hence, bomb-proofs are sometimes placed under the parapet; sometimes in the gorge of a half-closed work; sometimes

Fig. 43.

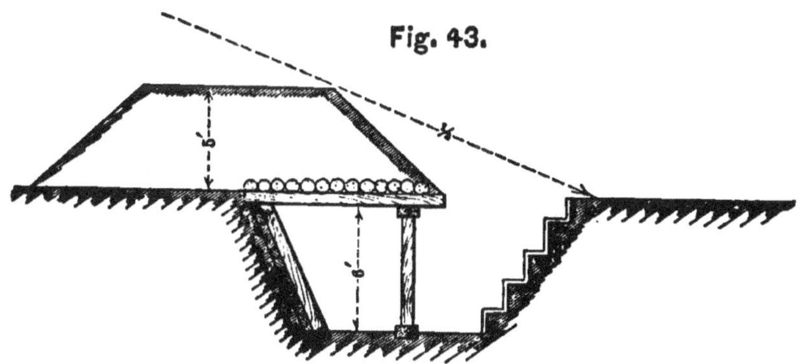

in the middle of the parade, etc.; the position being determined by the circumstances of each case.

Fig. 43 represents a cross section of a bomb-proof

into which the men can retire and be safe from the effects of a direct plunging or curved fire.

The construction of this bomb-proof was as follows: A row of vertical posts, in contact or at short distances apart, were framed into ground sills, or set into the ground at the bottom of the trench, along a line far enough away from the side of the trench next to the enemy's fire to allow room for a man to recline at full length, or occupy a comfortable position. These posts were then capped with stout timbers.

A second row of posts was placed, either vertical or inclined, against the side of the trench, and this row was capped in a similar manner. Cross timbers were laid in juxtaposition upon these capping timbers, and then covered with planks, fascines, etc., to form a tight roof.

This roof, when thus formed, was covered with sufficient earth to make it proof against the effects produced by the impact of projectiles, or the effects resulting from their explosion. A thickness of five feet of earth, in a vertical direction, is usually considered sufficient to make a roof proof against the effects produced by projectiles of field guns.

Ingress and egress of the men using the bomb-proof may be facilitated by cutting steps into the side of the trench, as shown in the figure. The part of the bomb-proof resting against the side of the trench should be revetted by a covering of plank, fascines, or other suitable material, to keep the shelter dry, and to make it more comfortable. Guard beds should be constructed, when the bomb-proof is wide enough, so that the men can lie down at full length; if not wide enough, benches can be made which will allow the men to assume easy positions.

A construction like that shown in Fig. 43, can easily be placed under the banquette, when the command is 9½ feet, the banquette in this case serving as the top of the bomb-proof.

114. Blindages.—Any construction used in field works which has for its object the protection of the men and material against the effects of artillery fire from overhead, is termed a **blindage.** The preceding construction, therefore, is a blindage.

115. Splinter-proofs.—Shelters which are not exposed to the impact of the projectiles of the enemy, need not be so strong as the bomb-proof. It will be sufficient if they are proof against the splinters and fragments of shells, produced by the enemy's fire.

Shelters of this kind are usually constructed in inclined positions. (Fig. 44). They are made by placing strong timbers, or bars of railroad iron, in an inclined position against the surface to be protected, and in juxtaposition, and then covering them with earth sufficient to make the interior safe against the fragments which may strike the shelter.

The inclination of the timbers will be equal to, or less, than the natural slope of the earth thrown against them. It is well to cover the pieces with raw-hides

Fig. 44.

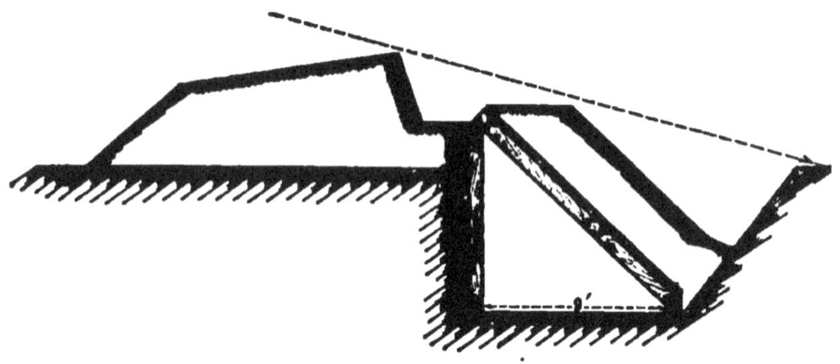

or tarpaulins before the earth is thrown against them, to make the shelter water-tight.

A thickness of two feet of earth is sufficient to resist the fragments of shells fired from field guns. In many cases the earth is placed upon the shelter by piling sand bags filled with earth against it.

Entrance to the shelter is provided for by openings at the ends, sometimes by openings left at intervals.

Splinter-proofs, from their nature, are placed in those situations where they are not exposed to a direct fire. They are much used to protect doors, entrances, etc., which are exposed to the effects of bursting shells; to protect vertical walls liable to injury from the same cause; etc.

116. Powder magazines, etc.—Shelters in

which the ammunition and other stores can be placed and kept safe from the effects of the enemy's fire, are equally as important as the shelters for the men. The most important of these are the **powder magazines**, or those shelters intended for the storage of the ammunition.

The rules given for the construction and location of bomb-proof shelters for men, apply equally to shelters of this class. The only difference in construction is in the size of the shelter, it being much smaller, as a rule, than that required for the use of troops.

Large magazines are not constructed in ordinary field works. They take up too much room, and even the best of them are but poor places in which to

Fig. 45.

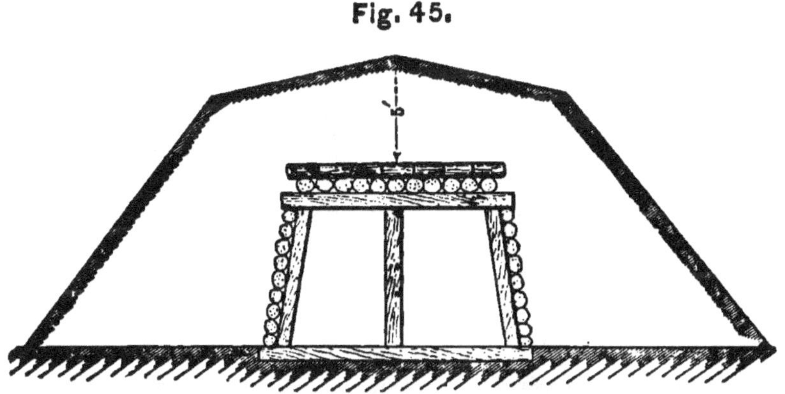

store ammunition for any length of time. The usual method adopted is to construct as many **service magazines** as may be necessary, near the guns to be served by them, making them large enough to contain the amount required for a definite service of the gun or guns to which they belong.

117. Service magazines.—Magazines of this kind are oftentimes built in the adjacent traverses (Fig. 45) if there be any; generally under the parapet near the guns; and sometimes under the barbettes.

Fig. 46.

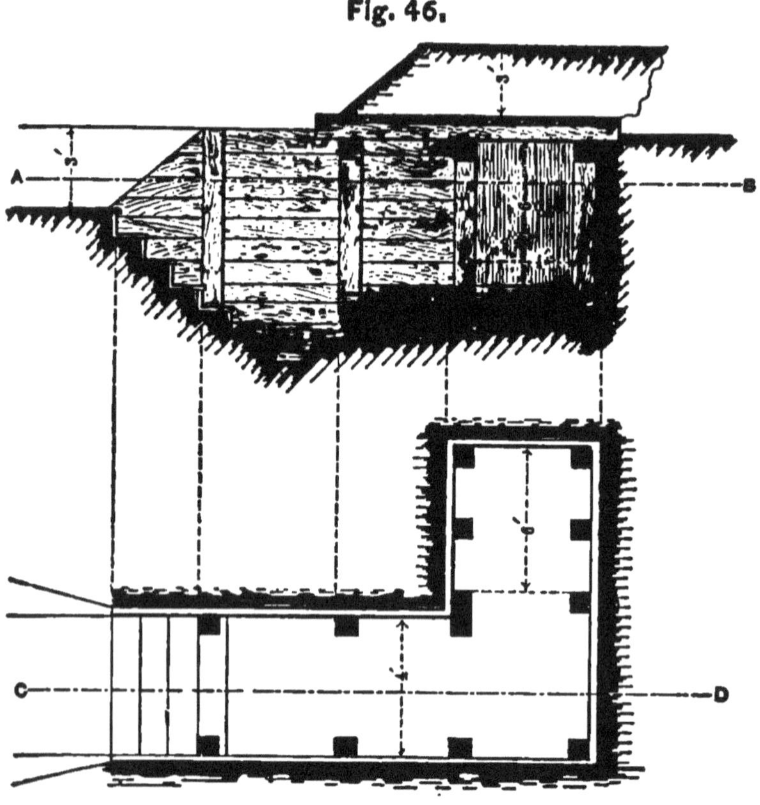

The conditions to be observed in locating and constructing a powder magazine are that it shall be conveniently placed; shall not be exposed to a direct fire of the enemy; be made bomb-proof; be well drained; and if practicable, be well ventilated.

A service magazine may be entirely above the surface of the ground (Fig. 45) or partly or wholly under the ground (Fig. 46).

The construction of a service magazine, in which the magazine and passage-way are lined with wooden frames, is represented in Fig. 46.

The **frames** are made of timbers or scantlings of the proper dimensions, each frame consisting of two uprights, called **stanchions,** a ground sill, and a cap. The interior dimensions of the frame are the same as that of the magazine, or six feet high and six feet wide, the least dimensions given, when practicable, to the width and height of the interior space.

The frames are placed upright, about three feet apart, and in the position which they are to occupy. Their tops and sides are then planked over; this planking is called the **sheeting.**

The bottom of the excavation is sloped from the sides to the middle, and from the rear to the front, to allow all water leaking through the magazine to collect in a shallow trench made along the middle line, and to run off into a drain prepared to receive it, or into a dry well dug near the entrance. The ground sills are then floored with boards.

Great care should be taken to make the top water-tight, before the earth is placed upon it. This done, it is covered with several feet of earth depending upon the degree of exposure to which it is subjected.

The plan, and horizontal section of the magazine and entrance, made by the plane A B; and the elevation, and section by the vertical plane, C D, are shown in Fig. 46.

The entrance to the magazine should be closed by a stout door, and the approach to it should be protected by a splinter-proof. If field artillery is employed to defend the work, the limber boxes are taken off and placed within the magazines.

118. Shelter for guns, etc.—Shelters are frequently provided for guns, implements, etc.

The thing to be sheltered, its dimensions, and its uses, will regulate the details of construction of the shelter. The rules applicable for the shelter just described, apply equally to shelters of this class.

119. Materials used in the construction of shelters.—Timber has been considered to be the material used in the construction of the above shelters. This material is so abundant in the United States that it can almost always be found in quantities near the work, and can be obtained quickly. It will therefore be the material chiefly used in temporary fortifications.

No better material can be used for the transverse pieces of these shelters than railroad iron, if it can be obtained. The form of the rails allows the pieces to be placed in juxtaposition without delay, and the strength of the iron makes the roof better able to resist the shocks of the projectiles, and makes the structure more durable in its character.

120. Bomb-proof shelters used in the defences of Washington.—The field works employed to defend Washington, in the war of 1861–5, are fine examples of temporary fortifications. They were in the beginning, constructed according to the rules laid down for temporary works. But as time passed, and their great importance was recognized, the dimensions of the parts of the different works were increased, and the interior arrangements so improved that the works passed beyond the limits laid down for field works, and approximated in profile and in the interior arrangements to permanent fortifications.

The section of a bomb-proof, shown in Fig. 47, gives a type of the shelters built for the men.

Fig. 47.

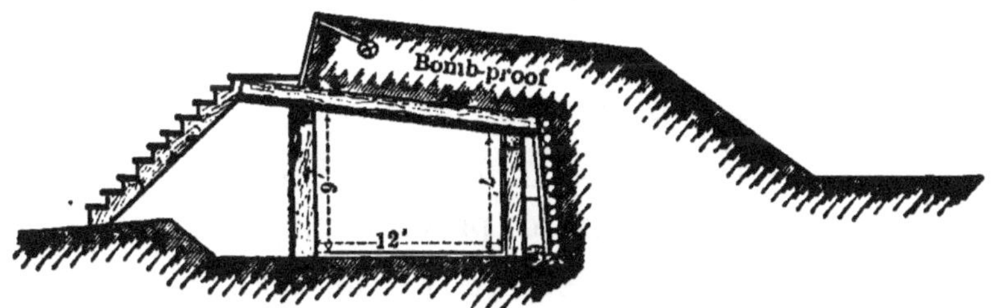

Its construction differs but slightly from those already described. It is larger, and is provided with a banquette by means of which a musketry fire could be delivered over the top of the shelter, if required.

The shelter represented by it was located near the middle of the parade.

121. Powder magazines used in the defences of Washington.—The powder magazines at first used were built like that described in art. 117. They were soon replaced by other constructions, of which the section shown in Fig. 48, represents the type.

Fig. 48.

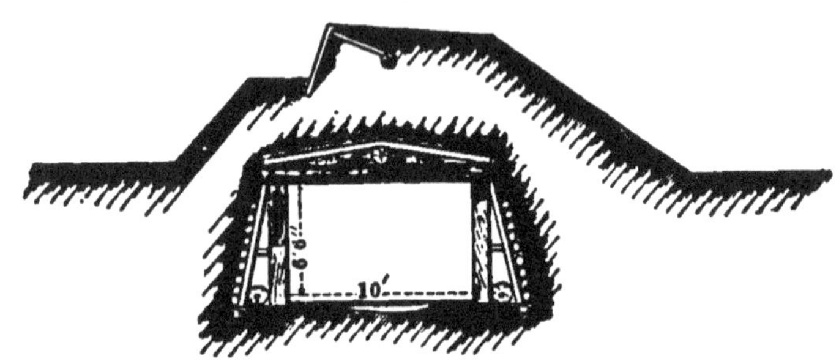

The new magazines were made stronger and more durable, were better drained, and were well ventilated; in all of which qualities the older ones were deficient.

This particular example shown in Fig. 48, was built under a traverse.

The essential difference between this particular magazine and those usually constructed, is the employment of an air-chamber to prevent the accumulation of moisture which otherwise would take place upon the inner side of the magazine. This chamber was constructed by placing logs in an inclined position against the sides of the magazine, framing them into a ground-sill, and bracing them at the middle. Small poles, from two to four inches in diameter, were laid against these inclined pieces, to hold up the earth thrown against them. Ventilators connected the interior of the magazine with the air chamber, and the air chamber with the outside.

Particular attention was paid to making the roof water-tight in all the magazines.

Whenever it was practicable, the drainage was carried to the main ditch of the work.

On the sides of a magazine exposed to the enemy's fire, the thickness of earth was made such as to measure at least ten feet from the woodwork of the magazine to the exterior, along a line making an angle of thirty degrees with the horizontal.

Magazines of the dimensions of those built in the defences of Washington, are not usually required, nor are they built in ordinary field works. But when a field work is to be occupied for some time, and a large quantity of ammunition is to be stored in temporary magazines, the principles governing the construction of the magazines should, as far as practicable, be in accordance with those illustrated by this example.

122. Position given to the shelters, etc.—No absolute rule, or set of rules, can be made which will apply to all cases in practice. Each particular work must be considered by itself and in connection with its surroundings. Interior arrangements, extremely necessary in one particular work, might be useless in another; the positions occupied by these arrangements in one work might be the worst places for them in a work of another kind, or in a work situated in a different locality.

The plan shown in Fig. 49, representing the

Fig. 49.

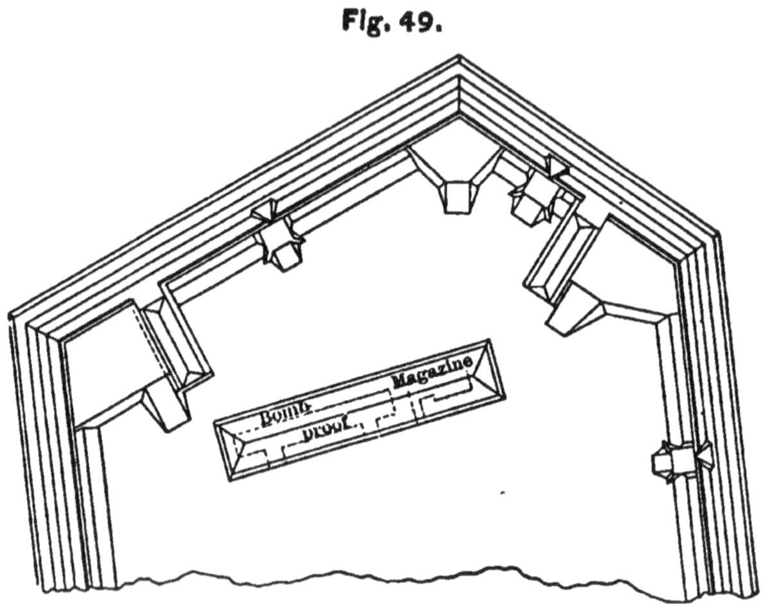

northern half of a redoubt used at the seige of Petersburg, in Virginia, in 1864, may be taken to illustrate the way in which the interior arrangements are sometimes located.

The trace of the redoubt was an irregular pentagon, and the portion shown in the figure was that part toward the enemy.

III. Communications, barriers, etc.

123. Communications.—The defenders of a closed work must have arrangements made by means of which they can enter or go out of the work when necessary. In case of continued lines, arrangements should be provided by means of which the defenders can make sorties.

The method adopted is to leave openings in the parapet, through which passages are built leading to the outside of the work. These openings made in a parapet are necessarily weak points of a work. As a consequence, they should be placed where they will be the least exposed, and can be most easily guarded. No greater number should be constructed than the actual necessities of the work demand.

In redoubts, the outlets are on the sides least exposed to attack; in half-closed works, they are placed near the middle of the gorge; in forts, they are usually placed near the re-entrants.

A passage for the use of infantry only should not, as a general thing, be less than six feet wide; for artillery, not less than ten feet wide; for sorties, the outlets in continued lines should be at least fifty yards wide.

124. Masks opposite the outlets.—The outlet should be masked in some way to prevent an enemy, on the same side of the work with the outlet, from seeing into the enclosed space. This is usually done by placing a traverse inside of the work, and opposite the outlet (Fig. 50).

Fig. 50.

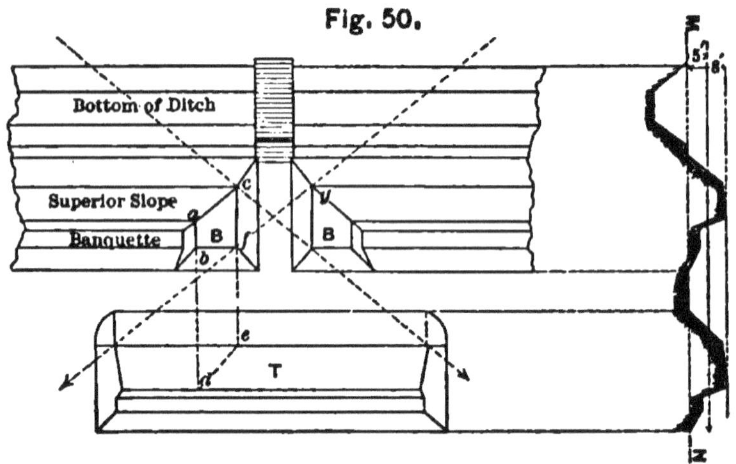

The traverse has, ordinarily, the same height as the parapet behind which it is placed, and a similar profile.

Room sufficient for a road-way, is left between the foot of the exterior slope of the traverse and the foot of the banquette slope of the parapet. The traverse must be long enough to intercept all projectiles which an enemy on the outside can fire through the opening in the parapet. Its length will therefore depend upon the width of the outlet and the thickness of the parapet.

The limiting plane of fire, above which a direct fire may be neglected, is taken to be five feet above the ground on which the man stands. The intersection of this plane of fire with the sides of the outlet should be determined, and the extreme lines of fire drawn. Let the dotted lines, through *c* and *g*, be the extreme lines of fire. If the exterior crest of the traverse rests on these two lines, the traverse will intercept all fire from the outside coming through the opening in the parapet.

When the height of the parapet is eight feet, it will be sufficiently accurate for all practical purposes to take the horizontal plane passing through the exterior crest as the limiting plane of fire, and to use the horizontal lines passing through *c* and *g*, shown dotted in the figure, as the extreme lines of fire.

The more accurate method would be to determine these extreme lines, by passing a plane five feet above the ground, and finding its intersection with the sides of the outlet, and with the exterior and interior slopes of the parapet. Join, by a straight line, the point of intersection of the line cut from the exterior slope and either of the lines cut from the sides of the outlet, with the point of intersection of the line cut from the opposite side of the outlet and the line cut from the interior slope. This line, thus drawn, will be one of the limiting lines. In a similar way, the other line could be obtained.

125. The length of the traverse may be shortened by turning back the interior crest at right angles to its general direction, and extending it as far as the crest of the banquette (B B, Fig. 50).

Instead of having a road along its entire front, the **traverse** is sometimes joined to the parapet on one side of the opening, as shown by the dotted lines *b d* and *e f*, in Fig. 50.

Sometimes, it is necessary to dispense with a traverse in the interior. Especially is this the case where the outlet is a very wide one, and the interior space quite limited.

The method adopted to mask the interior of the work in this latter case, is to place the traverse opposite the outlet on the outside, and beyond the ditch. (Fig. 51).

Fig. 51.

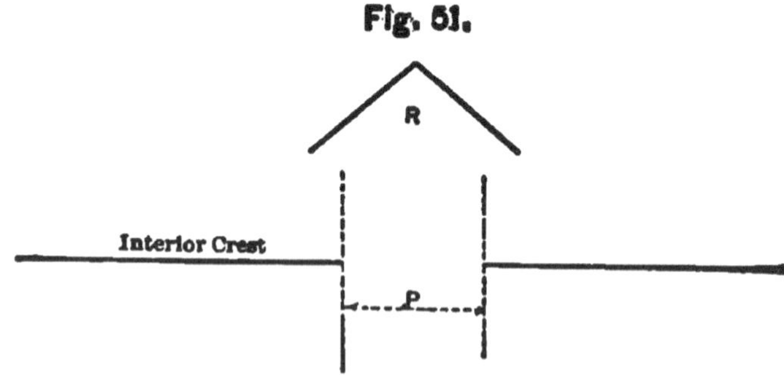

The traverse in this case is usually broken, generally a redan in trace, with the profile of a parapet, but commanded by the parapet in rear.

126. Barriers.—The outlets are usually arranged so that they can be quickly closed, to guard against surprise. The means used is a gate, technically termed a **barrier**.

The gate is made with two leaves, hanging on posts by hinges, and made to open inward.

The frame of each leaf is composed of two uprights, called **stiles**; two cross pieces, one at the top and the other at the bottom, called **rails**; and a diagonal brace, called a **swinging bar**.

The leaf of the barrier may be made open, by spiking stout upright pieces, with intervals between them, to the pieces of the frame; or it may be made solid, forming what is known as a **bullet-proof** gate. (Fig. 52).

Fig. 52.

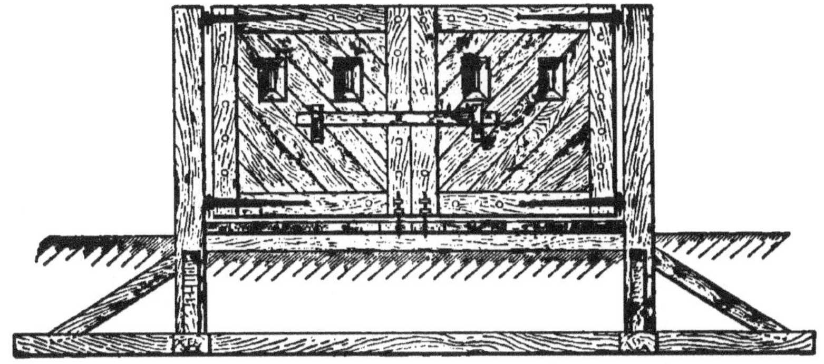

Since the gate must be strong, the leaves of it are necessarily very heavy. The leaves must be hung upon stout posts, firmly braced into the ground, to sustain the great weight of the gate.

The top rails of all barriers should not be less than six feet above the ground.

In the barriers with open leaves, the vertical pieces are usually extended from eighteen inches to two feet above the top rails, and their upper ends sharpened.

In those which are solid, it is usual to arrange some obstruction upon the top rail, such as sharp pointed spikes, broken glass, etc., to interfere with persons climbing over the top. It is usual to provide apertures in the leaves, through which the men can fire upon the ground on the outside.

127. Bridges.—When the ditch has been completed along that part of the work in front of the outlet, it is usual to carry the roadway across the ditch by means of a bridge.

The ditches of field works are, as a rule, quite narrow, and the bridges used to span them are very simple constructions.

A common method of building the bridge is to lay three or more sleepers across the ditch, and cover them with planks laid transversely. If the span is sufficient to require intermediate supports, these are obtained by using trestles placed in the ditch.

A bridge built in this way can be quickly removed and speedily re-built, if there be any necessity for it.

Hand-books on military engineering describe a number of bridge constructions for use in fortifications. These bridges are all arranged so as to admit of communication across the ditch being interrupted at pleasure, and are either draw or rolling bridges.

The draw-bridges are usually made with a leaf to revolve around a horizontal axis, the leaf usually being raised to a vertical position when the communication is interrupted. The rolling bridges are arranged to be pushed out from the work, and drawn back into it. Bridges of this class, known as movable bridges, are useful to guard against surprise, to prevent stragglers from entering, and to keep the garrison in the work. As a defence against an assault of a field work, they are of but little value.

The best method, is *to have no ditch* in front of an outlet, but let the roadway be on the natural surface of the ground.

128. Ramps.—The short roads used in fortifications to ascend from one level to another, are termed **ramps.**

The width of a ramp depends upon its use, following the rule laid down for the width of passages. A width of six feet for infantry, and of ten feet for artillery, are the widths generally used.

The inclination of the ramp may be as great as one on six, and as little as one on fifteen, depending upon the difference of level between the top and bottom. The side slopes are of earth with its natural slope.

The ramps in a work should be placed in positions where they will not be in the way, nor occupy room which may be required for other purposes.

Steps or stairways are sometimes used instead of ramps. The rule for them is that the breadth of each step, called the **tread**, shall be at least twelve inches, and the height of the step, known as the **rise**, shall be about eight inches.

They are substituted for ramps in those places where there is not sufficient room for the ramp.

IV. Arrangements Intended for the Comfort and Health of a Garrison.

129. Nature of the arrangements.—A garrison compelled to live within an enclosed space like a field work, should be provided with all the arrangements which are necessary for the health and for the comfort of the men, consistent with surrounding circumstances.

The arrangements essential to the health and comfort of the men include those intended to protect them from the weather, to provide for their support, and to supply their necessities.

The principal arrangements are the tents, huts, or shelters in which the men are sheltered; the guard-houses, and rooms for those on duty; the kitchens and bake-ovens in which the food is prepared; the sinks or privies, and the places provided for the men for washing themselves and their clothing; the hospitals for the sick; etc. Wells, or means of providing the garrison with a supply of good drinking water, form no unimportant part of the arrangements necessary for the comfort as well as the health of a garrison.

The limits of this book will not admit of a discussion, nor even a reference to the various divisions, of this important section of the interior arrangements of a field work.

These arrangements are second only to those required for actual defence, and in many cases they are equal, as the defence of the work in a great measure depends upon them.

The only rule that can be laid down is to make all these arrangements of a temporary character, and to place them so that they can be removed, at a moment's notice, out of the way of any interference with an active defence of the fortification.

V. Other kinds of Interior Arrangements.

130. Secondary interior arrangements.—Besides the interior arrangements which have been described, or mentioned, there are others which are secondary in their nature. These are the arrangements which are to be used under certain contingencies, or in cases of emergency. An example, would be a defence placed within a field work, which defence can be used only when the main work is no longer defensible, etc.

131. Block-houses.—It is frequently the case that a separate fortification is constructed, lying entirely within a work exterior to it, into which a garrison can retire and protract their resistance, even after the outer fortification has been taken, or has been made unfit for further defence.

If this interior work is a line of earthern parapet, it is called a **retrenchment**; if it is a defensible building, it is termed a **keep.**

The term, keep, is also applied to a work which is entirely separate and distinct from the work exterior to it, whatever may be the material used in its construction. In a field work, the keep is built of timber, and is called a **block-house.**

The conditions which should be fulfilled by a block-house intended for a keep, are that it shall have a good command over all of the interior space enclosed by the outer work, and shall occupy a position such that all parts of the exterior work can be seen from it.

The plan of a block-house is selected by the same general rules which are used for selecting the trace of a field work. It may be square, rectangular, octagonal, and even cruciform, in plan, according to the locality in which it is placed and the fire which it has to deliver.

The dimensions of a block-house should be sufficient to allow sleeping accommodations for the men who are to occupy it; and in some cases allowance should be made for other accommodations. Its interior dimensions should give at least a height of six feet in the clear for the rooms; a height of eight or nine feet gives better accommodations and better ventilation.

The width of the interior should not be less than nine feet in the clear, as this is the least distance which can be used and give room for a passageway and a row of bunks.

The length of a block-house will depend upon the number of men it has to accommodate, after the width has been assumed.

Block-houses must be made strong enough to resist the projectiles which may strike them and should be proof against fire and splinters.

They should be free from dampness, and should be well ventilated.

The conditions given for a bomb-proof are applicable to the block-house, with the additional one of arranging its walls for defence.

This is fulfilled by perforating the walls with loop-holes for musketry and embrasures for cannon, when the latter are used.

The example given in Figs. 53, 54, and 55, shows the details of construction of a wooden block-house, used in the late war in the United States.

The walls were made of two thicknesses of logs (Fig. 53) the inner row being vertical, the outer row being horizontal.

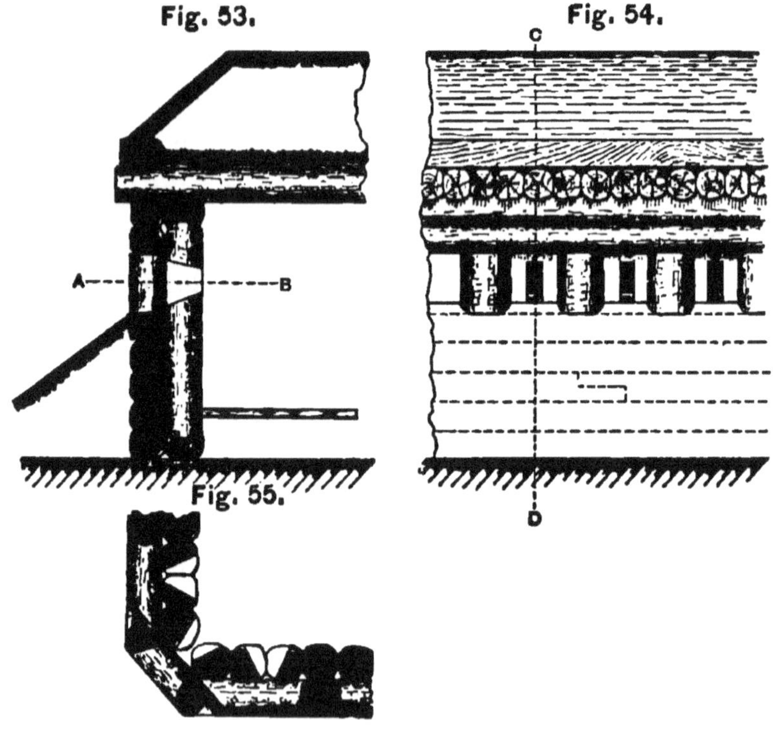

Fig. 53.

Fig. 54.

Fig. 55.

The inner row was composed of logs in contact, framed into a ground-sill, and capped by a heavy piece of timber. The logs in the outer row were laid one upon the other, having a surface of contact of at least eight inches.

A roof of logs was laid upon these rows forming the walls, and then boarded over to make it water-tight.

This roof was covered with earth from three to four feet deep, to make it fire and splinter proof.

The sides exposed to the enemy were banked with earth (Fig. 53) to increase their resistance to the enemy's projectiles.

Loop-holes were constructed through which the defence could fire upon the ground exterior to the block-house.

132. Isolated block-houses.—Timber block-houses were used frequently in the war of 1861–5 in isolated spots, as independent works.

In these places, they were, as a rule, exposed to attack only from infantry or cavalry, or a few pieces of field artillery.

The construction shown in Figures 53, 54, and 55 is a type of these isolated block-houses.

It was found from experience that it required a thickness of forty inches of solid timber to resist the projectiles of field-guns.

These isolated block-houses were frequently built two stories high. The upper story was usually placed so as to have its sides make an angle with the sides of the lower story. By this arrangement, the corners of the upper story projected over the sides of the lower. This arrangement of the upper story removed the dead space near the sides of the lower story, and the sector without fire in front of the angles. Block-houses exposed to artillery fire should not have a second story.

133. Stockades.—A line of stout posts or trunks of trees firmly set in the ground, in contact with each other, and arranged for defence, is called **a stockade.**

A stockade is used principally when there is plenty of timber and little or no danger of exposure to artillery fire. It is frequently used to close the gorge of a field work, and to guard against the work being carried by a surprise, by bodies of infantry attacking the work in rear.

The timbers of a stockade may be either round or square. If round, they are hewed to a flat surface on two of the sides so that the posts, when placed in position, shall have a close contact of at least four inches.

The top of a stockade should be at least eight feet

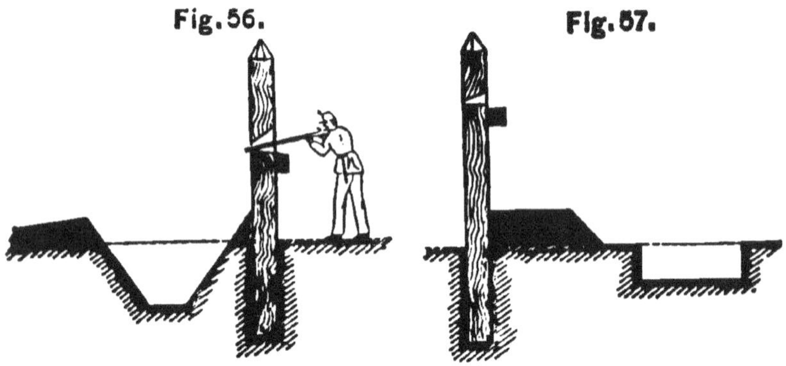
Fig. 56. Fig. 57.

above the ground on which it is placed, and it should have the upper ends of the timbers sharpened, or arranged with spikes, or fixed in some way to offer an obstruction to climbing over the top. A stockade is arranged for defence by cutting loopholes (Figures 56, and 57,) which can be used by a soldier when firing.

The height of the loop-hole may be just four feet and six inches above the ground, or higher.

In the former case, the soldier stands upon the natural surface of the ground (Fig. 56) in the act of firing; in the latter case, he stands upon a banquette of earth (Fig. 57) or some temporary arrangement which raises him to the proper height.

The exterior opening of the loop-hole should be not less than six feet above the ground on which the enemy may stand when he is close to the stockade, so that he can not make use of the loop-hole. Some obstruction must therefore be placed in front of the

Fig. 58.

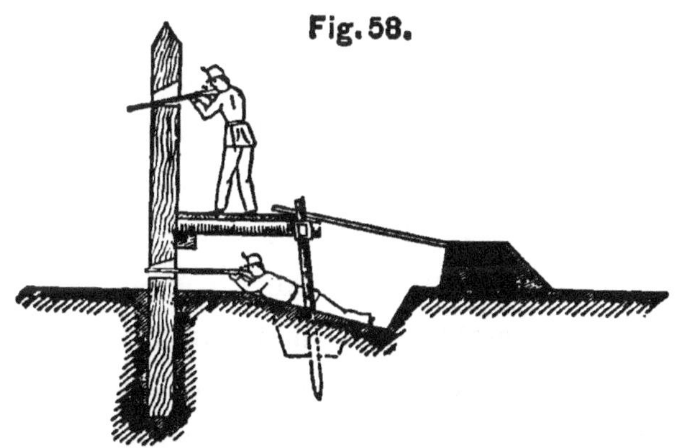

loop-hole, keeping the enemy away from it; or the ground immediately in its front should be deepened by digging a trench (Fig. 56).

If the loop-hole can be placed six feet above the

ground, it will be practicable to arrange the loop-holes so as to furnish a double tier of fire (Fig. 58).

The loop-holes of the lower row should be arranged so that they can not be used by the enemy. They should not be higher than eighteen inches above the ground on the outside; on the inside, the ground must be cut away (Fig. 58) or a trench dug in rear of the stockade, so as to allow the use of the loop-holes by the defence.

The construction of a loop-hole in a stockade can be understood by examining Fig. 59.

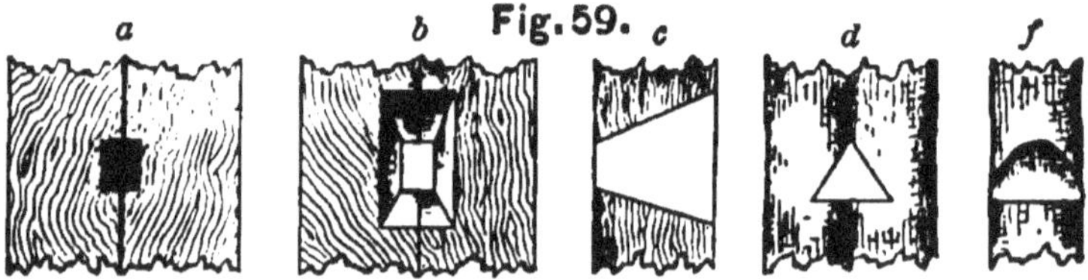

Fig. 59.

The exterior opening of a loop-hole is represented by *a*; the interior opening by *b*. A side view is represented by the longitudinal section *c*.

The same rules govern the construction of loop-holes when there is time to execute them, as are used in constructing embrasures in a parapet.

The exterior opening is made as small as possible and serve its purpose. An opening of two inches wide and five inches long is considered sufficiently large for the rifled musket of the present day (Art. 108).

The size of the interior opening will depend upon the field of fire. A width of six inches and a height of twelve inches are the dimensions usually given to the interior opening, when the timbers of the stockade are twelve inches thick.

When there is a great hurry, a notch may be cut by an axe or a saw in each of the logs, which will do for a loop-hole. The front of this notch is shown by the figure *d*, and the side view by the figure *f*.

A perpendicular cut of four and a half inches is made in each log at the point which is to be the bottom of the loop-hole. An oblique cut is then made about nine inches long. The two logs placed in contact form a loop-hole as seen in *d* (Fig. 59).

The exterior opening may be reduced afterwards by spiking a piece of iron upon the logs. Railroad

Fig. 60.

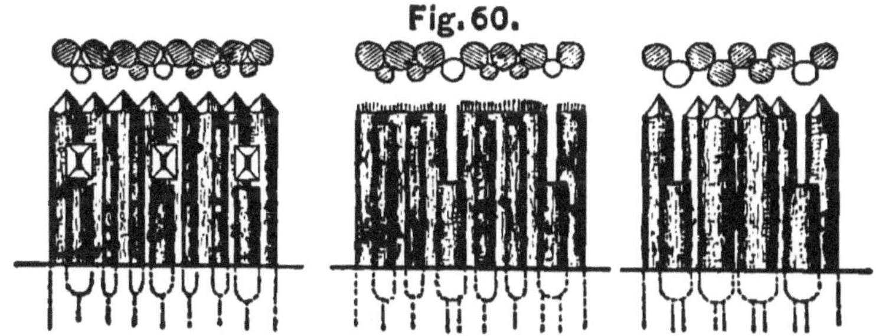

chairs, fish-plates, etc., frequently found in the vicinity, can be used for this purpose.

Sometimes the stockade is strengthened by a second row of timbers. The method is shown in Fig 60.

This figure shows also the method of making loop-holes, by cutting away a strip from two adjacent posts, leaving an interval through which the men can fire.

Loop-holes are usually made about two feet and six inches apart.

CHAPTER XII.

ARRANGEMENTS EXTERIOR TO THE PARAPET.

134. Kinds.—There are two kinds of arrangements employed on the exterior of a parapet to add to the strength of a work, viz: the arrangements made to defend the ditch, and those made to obstruct an enemy's approach.

The term, **ditch defence**, is used to designate the arrangement made exterior to the parapet, by which a fire is made to sweep the ditch.

The term, **obstacle**, is applied to any construction or arrangement, whatever may be its nature which, by its passive resistance, obstructs the approach of an enemy advancing to assault the work. Hence, anything is an obstacle which diverts the attention of the enemy from the assault to the immediate surroundings of himself.

I. Ditch defences.

135. Classes of ditch defences.—The ditch is best defended, as a general rule, by the work itself. But as a flanking disposition is not always attainable, and as it is not usually practicable in field works, some arrangement must be provided by means of which the ditches can be swept, in those works where it is necessary that the dead spaces should be removed.

The constructions used for this purpose are either **caponnières** or **galleries.**

136. Caponnières.—A caponnière is a bombproof construction built in the ditch (Fig 61).

Fig. 61.

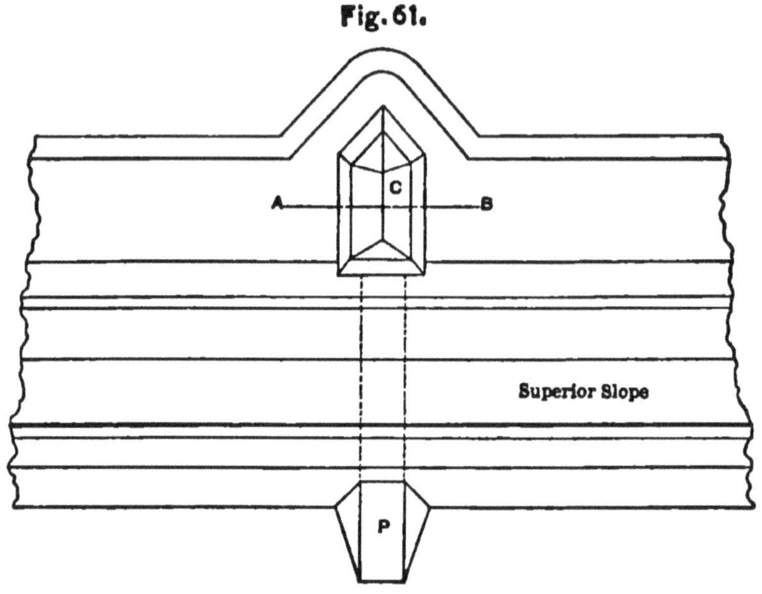

It may be arranged for a fire in two directions, or only in one. The amount of fire fixes its length.

When arranged for a fire in two directions, it should be wide enough to accommodate two ranks, facing in opposite ways. This width should not be less than eight feet. A width of four feet and six inches is sufficient for one rank.

The interior height should not be less than six

feet in the clear (Fig. 62). The bottom of the caponnière may be on the same level, as the ditch, or below it.

Fig. 62.

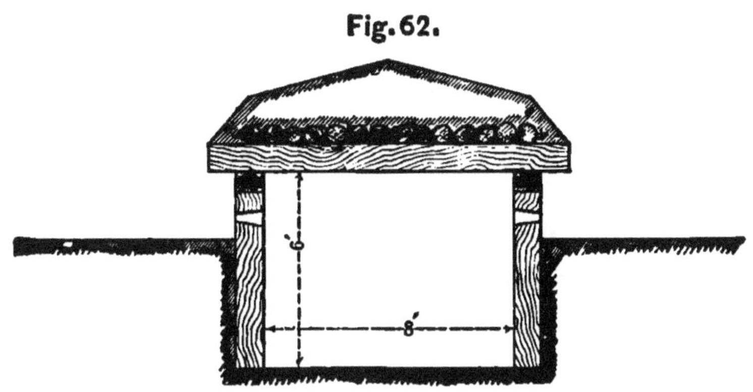

The loop-holes are arranged as described in art. 133. The top should not rise above the crest of the counterscarp and should be made bomb-proof.

Entrance is obtained by a covered passage, P, leading under the parapet. It is constructed by using frames like those described in Art. 117.

The construction of the roof and the position of the loop-holes are shown in Fig. 62, which represents a cross section made by the plane A B, in Fig 61.

137. Galleries.—Galleries used to sweep the ditch with their fires are constructions similar to caponnières. They receive the name of **scarp** or **counterscarp** galleries, according to the position which they occupy.

The usual method of building a gallery is to excavate the earth behind the scarp or counterscarp, as the case may be, until there is room to place the frame, if there is one to be used, or until there is room to accommodate the men who are to occupy the gallery. This excavation is then closed in front by a stockade built along the line of the scarp or counterscarp; it is closed overhead by a bomb-proof roof, the top of which is the parapet, in the case of a scarp gallery, and the glacis or natural surface of the ground, if a counterscarp gallery.

The arrangements of loop-holes, dimensions, etc., are all similar in kind and character to those mentioned for caponnières.

Entrance to a scarp gallery is by a covered passage; the same might be used for a counterscarp gallery; generally, the entrance to the latter is by openings into the ditch, at the ends of the gallery, which openings can be closed by bullet-proof doors. Counterscarp galleries are used at the salients of a work; scarp galleries at the re-entrants.

Scarp and counterscarp galleries, as well as caponnières, are, in their details of construction, nothing but stockades. It follows then that all remarks relating to stockades apply equally to these ditch defences.

They should not be placed in positions where they would be exposed to artillery fire. If there was danger of exposure to this fire, and it was necessary to use one of these defences, it is plain that the counterscarp gallery is to be preferred.

It may be remarked, however, that ditch defences of this kind rarely repay, by their advantages, the amount of time and labor expended in their construction; and then, only in the case of a large enclosed work with deep ditches.

II. Obstacles.

138. Kinds.—Obstacles are of two general kinds, viz: **natural,** and **artificial.**

Marshes, water courses, precipices, etc., are examples of obstacles of the first kind. A very little labor, well applied, will often convert these natural features of the ground into serious obstacles to an enemy's advance.

Ditches, abatis, palisades, slashings, etc., are examples of artificial obstacles. They can be made important adjuncts of a defensive work, and can be made to perform effective service in the defence of any position.

139. Ditches.—The ditches, from which the earth is obtained to make the parapet, can be made into obstacles to an enemy's assault. To offer an obstruction difficult to surmount, a ditch should be not less than six feet deep and twelve feet wide. (Art. 28). The difficulties of surmounting an obstruction of this kind can be greatly increased by making the scarp steep, and by a judicious arrangement of some of the other obstacles, hereinafter named, in and along the ditch.

When the scarp is made steep, it is necessary to revet its surface, so as to protect it from the weather and to hold the earth in place. In the late war in the United States, the revetments used for the scarps of works when strength was required, were made of timber or plank.

The **timber** scarp revetment (Fig. 63) was com-

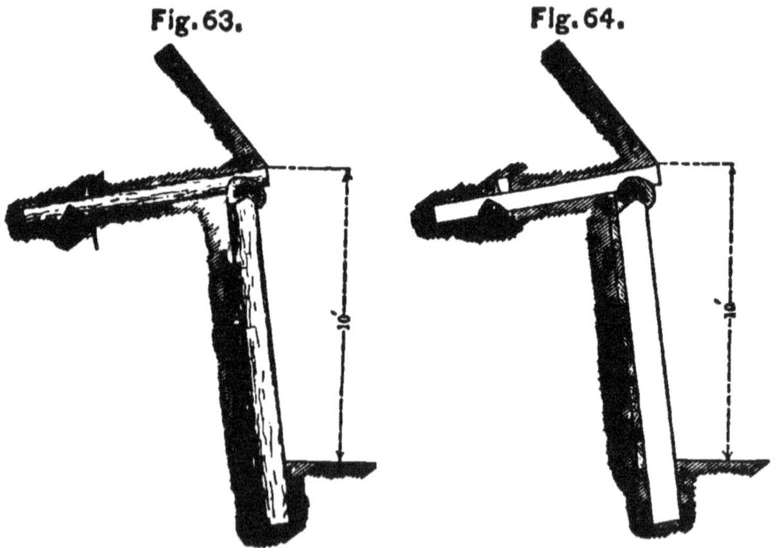

posed of logs placed in contact with each other, with a slight slope, about $\frac{1}{12}$, and capped by a log notched on the uprights; this capping log being tied back to an anchor log, bedded into the natural soil and held in place by pickets.

The **plank** scarp revetment (Fig. 64) differed but

slightly from the timber revetment. The posts, instead of being in contact with each other, were placed from six to eight feet apart, and the earth was held up by planks placed on edge behind the posts.

These revetments are practically the same as the revetments described in articles 81 and 82.

The timber revetment is the stronger of the two, but requires more time, labor, and timber to construct it than the plank revetment.

140. Abatis.—A row of large limbs and branches of trees, with the ends of the branches sharpened, placed with the points towards the enemy, forms the obstacle known as an **abatis.**

An abatis may occupy an upright position in a ditch (Fig. 65) or it may lie in a horizontal position

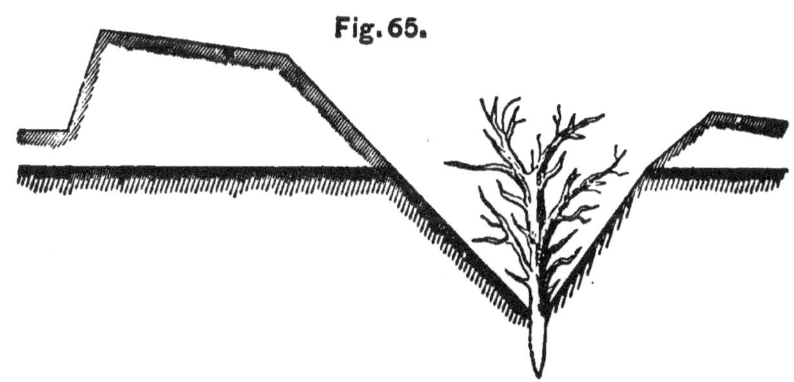
Fig. 65.

on the ground in front of a work (Fig. 66). The latter is the usual position given to an abatis.

An abatis, to be an efficient obstacle, should consist of stout limbs, twelve or 15 feet long, laid as close together as possible, with the butts secured by stout stakes, and should not be exposed to the enemy's artillery fire. When laid in front of a work, it is usual to protect it by a slight glacis. (Fig. 66.)

Fig. 66.

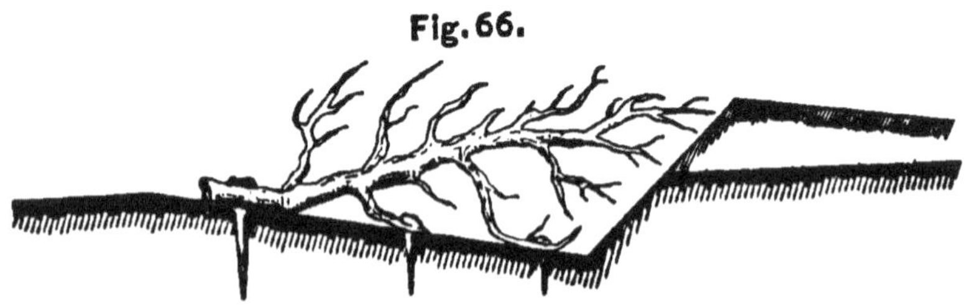

141. Entanglement.—An entanglement made by driving stout stakes into the ground from six to eight feet apart and connecting them by stout wire twisted around the stakes, forms an excellent obstacle. (Fig. 67.) It is quickly made when the materials are close at hand.

Fig. 67.

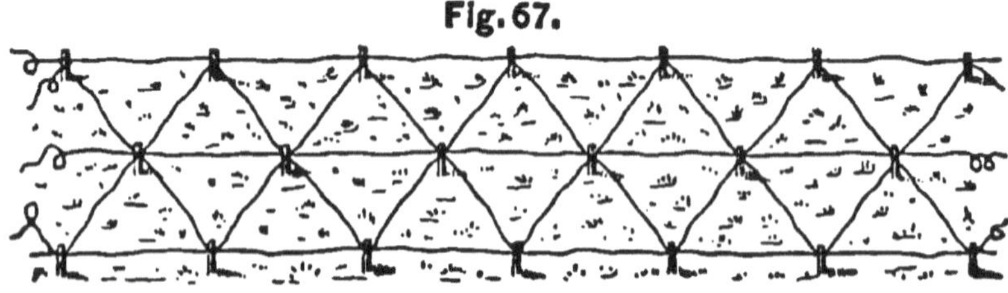

142. Chevaux-de-frise.—A cheval-de-frise (Fig. 68) is a square timber, perforated with holes in which sticks shod with iron are inserted. The holes are placed about six inches apart, and are large enough to admit a stick of at least two inches in diameter.

These sticks, called **lances**, are ten feet long, and **are** made to project equally on the two sides of the **body,** as the timber is called. The length of each cheval-

Fig. 68.

de-frise is from six to ten feet. They are fastened together at the ends by stout wire, or by chains.

In the British service, portable chevaux-de-frise made of iron, and in lengths of six feet, are used.

143. Palisades.—A palisading is simply a fence,

Fig. 69.

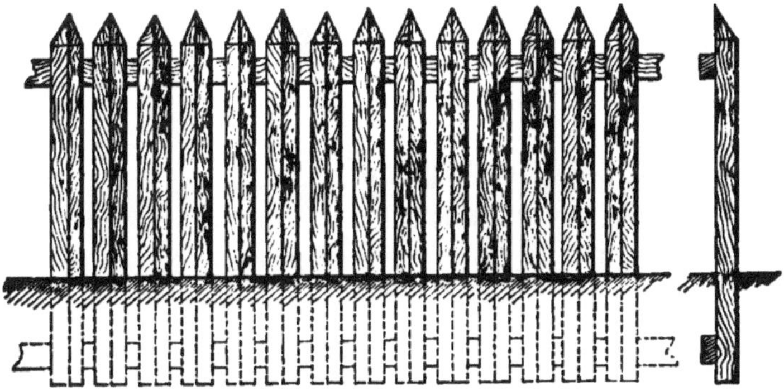

made of strong and stout poles or pickets firmly set into the ground (Fig. 69).

The poles are placed in a vertical or slightly inclined position, with intervals between them of about three inches.

The poles are obtained by sawing trunks of trees into lengths of from ten to twelve feet, and splitting them into rails of not less than four inches thick.

The palisade is formed by digging a trench three feet deep, planting the rails in it, and ramming the earth around them.

The rails are fastened together by being spiked or nailed to a stout **riband** piece placed about a foot below the surface of the ground. A second riband piece is used about a foot below the top of the rails.

It is usual to sharpen the rails to a point to add to the difficulties of climbing over the palisade.

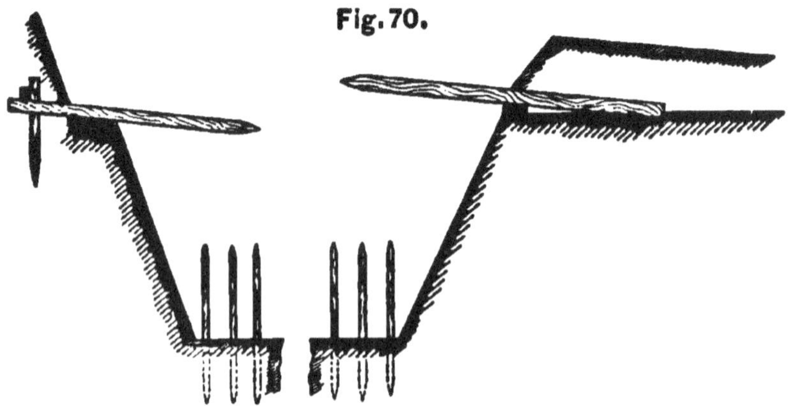
Fig. 70.

144. Fraises.—A palisading occupying a horizontal position (Fig. 70) is called a **fraise**.

145. Small pickets.—Small straight branches of hard and tough wood are frequently cut into short lengths, driven firmly into the ground, leaving from one to two feet projecting, and pointed at the upper end. Obstructions of this kind are called small **pickets** (Fig. 70).

146. Crow's feet, harrows, etc.—Obstacles of other kinds having the same object in view as small pickets are sometimes used. The most noted are the obstacles known as **crow's feet**; ordinary harrows turned upside down with the teeth upwards and the frames buried; planks with spikes driven into them, placed so as to have the points upwards; etc.

A crow's foot is made of four stout iron spikes, welded together at their heads, and arranged so that in whatever position the obstacle is thrown upon the ground, one of its points shall stand upwards.

147. Military pits.—Excavations made in the ground, conical or pyramidal in form, with a small picket driven into the bottom, are called **military pits.** (French, *trous-de-loup.*)

They are of two kinds, viz: deep and shallow.

Deep military pits (Fig. 71) should not be less than six feet in depth, so that if they fall into the possession of the enemy, they can not be used against the defence.

They are usually made about six feet in diameter at top, and about one foot at the bottom, and are placed so that the centres shall be about ten feet apart. They should be placed in rows, at least three in number, the pits being in quincunx order. The earth obtained by the excavation, should be heaped up on the ground between the pits.

Fig. 71.

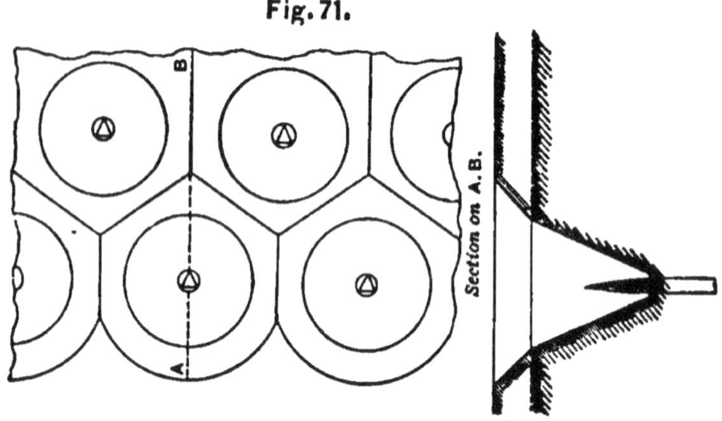

Shallow pits should not be deeper than about two feet, so that the enemy could not obtain shelter by getting into them.

They should cover the ground in a zig-zag arrangement, the upper bases being made square or rectangular in form, and in contact with each other. The side of the upper base should be made about equal to the depth of the pit. The earth obtained from the holes is thrown in front of the arrangement, making a glacis.

148. Slashing.—In compliance with the principle that all houses, trees, brushwood, etc., within range of the work, which could be used as a shelter and a place of concealment by the enemy's sharpshooters, should be removed, it is essential that the trees within six hundred yards of the work should be cut down.

As it is not practicable to remove immediately the trees from the spot, it is the custom to cut them down so that they shall form, while lying on the ground, an obstacle which may be used in the defence of the work.

Trees cut down so as to fall in all directions, form what is known as a **slashing.** It is better, where the trees are intended to be used as an obstacle, that they be cut so as to fall towards the enemy; and, in the case of the smaller trees, which might be moved by a few men, the trunks should not be cut entirely through, but only enough to allow the trees to fall. (Fig. 72.)

Fig. 72.

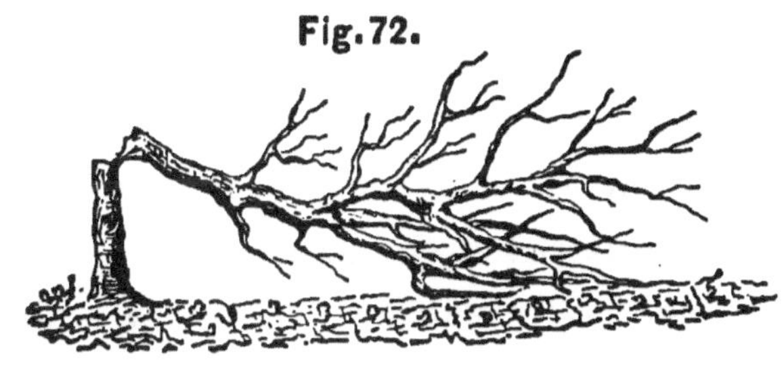

A thick and well arranged slashing forms an excellent obstruction to an enemy's free movements. It has the serious defect of being easily burned when dry.

149. Torpedoes.—Loaded shells buried in the earth just deep enough to be concealed, and arranged so that they can be exploded automatically, or at the will of the defence, have been used as obstacles. Arrangements of this kind are known as **torpedoes.**

The case enclosing the charge may be either of wood or iron. Condemned shells are especially suitable for the purpose.

The explosive compound used to charge them may be powder, gun-cotton, nitro-glycerine, or any material which, upon being fired, will burst the case containing the charge and scatter the fragments in every direction.

The automatic—sometimes known as the "sensitive" torpedo—is fired by contact. It has the advantage of being exploded at the right time, but has the disadvantages of making the ground, in which it is buried, dangerous to the defence, and of subjecting the men when handling it to the danger of accidental explosions.

The torpedo which is fired "at will" has the disadvantage of being fired oftentimes prematurely, or when it is too late.

Circumstances can only decide as to which of the two is to be preferred as an obstacle.

150. Stone-fougasses.—Military mining is rarely resorted to either in the attack or the defence of a field work. When used, it forms part of the operations of a siege, which may sometimes be under taken to get possession of a field work.

A mine placed at the bottom of a small shaft and fired by a fuze, is termed a **fougasse.**

A particular arrangement known as the **stone fougasse** is always described by military writers on the subject of field works as a useful device in the defence of a work.

It is usually constructed by excavating a funnel shaped hole in the ground to a depth of five or six feet. The axis of the excavation has an inclination of about 40° with the horizon ; the lower line has an inclination of about 30°.

A box, containing the powder, is placed in the bottom of the hole, and covered by a wooden shield which exactly fits the excavation. The earth of the excavation is well rammed around the shield on top and behind, so as to make the line of least resistance offered to the explosion coincide as nearly as possible with the axis of the excavation. On this shield and in this hole are poured from three to five cubic yards of stones, the smallest weighing one pound. Sometimes loaded shells are mingled with the stones.

The fougasse is fired by means of a fuze, or by electricity, in a similar manner to that used for firing mines.

The result of the explosion is to scatter the stones in a shower over a considerable extent of ground in front of the fougasse.

The whole construction is simply an extemporized

mortar for throwing stones, capable of a single discharge.

Its position must be concealed from the enemy's view.

151. Shell-fougasses.—A shell fougasse is a box containing loaded shells, concealed in the earth, and so arranged as to be exploded when the enemy is over the spot.

The box is divided by a partition into two parts, an upper and a lower. The loaded shells are placed in the upper part, with the fuzes downwards and connecting with the lower part by holes bored in the partition.

A charge of powder is placed in the lower division of the box of sufficient quantity, when fired, to throw the shells to the surface. This charge is fired by means of a fuze, or by electricity, like other fougasses.

152. Inundations.—If the position occupied by a field work is near a stream, it may be possible, in some cases, to increase the difficulties of the enemy's approach by inundating the ground over which he has to pass. The **inundation** may be produced by building dams and causing the waters of the stream to overflow its banks.

If the depth of the water over the approaches is greater than five feet, the obstacle may be considered as practically insurmountable.

If the depth is less, the obstacle is still a serious

one, and can be made greater by digging pits and ditches at random and having them covered with a sheet of water. The ditches of the work, already important obstacles in themselves, can, by flooding them, be made almost impassable.

153. Uses of obstacles.—No obstacle is insurmountable. Obstacles may hinder, and even stop for a while, an enemy's approach, but they can be overcome. Their passive resistance must be aided by the active resistance of the defence.

These obstacles, in order that they should be *accessory* means of defence, should detain the enemy in a position where he will be under the fire of the defenders at close range. Hence, the following conditions should be observed in arranging the obstacles in front of a field work.

1. The obstacles should be placed within close musketry range of the defence.

2. They should be arranged so as not to afford shelter to the enemy.

3. They should, as a rule, be protected from the fire of the enemy's artillery.

4. They should be arranged so as not to interfere with an active defence of the work.

Their uses as obstacles will depend upon the degree of resistance which they offer in harmony with the foregoing conditions.

Abatis placed in the ditch will, in one case, be in

the best position; in another, it should be placed some distance in front of the work.

A fraise placed in the scarp, when the ditch is swept by a fire from the work or from ditch defences, will be better than if along the counterscarp.

Torpedoes, military pits, entanglements, etc., may all be combined. In some cases, the ground in front of the work will be the better position; in others, the crest of the counterscarp and the ditches offer the best conditions for their use.

As a general rule, it is advisable to place the obstacles not nearer than fifty yards to the interior crest, if the profile is a weak one. When the profile is strong, it is not a matter of so much importance, so long as the assaulting columns are exposed to the fire of the defence.

It is well to remark with respect to inundations, that they should not be used until the last moment. The unhealthiness due to the presence of stagnant water is apt to produce more casualties than are to be feared from the enemy's attacks.

If the dams can not be protected or hidden from the enemy's artillery fire, they should be built, as far as possible, so that the enemy can bring his fire to bear only upon the upper side. The amount of the dam exposed to his fire will then only be the portion between the top and the surface of the water.

CHAPTER XIII.

FIELD FORTIFICATIONS UPON IRREGULAR SITES.

154. General considerations.—In the preceding chapters, the site of the field work and the ground in its immediate vicinity were considered as practically level.

The case now to be considered is the one in which the site and the ground in the immediate vicinity of the work are irregular in formation.

It has been seen how the trace and the profile of a work are modified by the presence of neighboring heights, which command the work. (Chap. X.) It is now to be shown how the irregularities of the site modify the profile and trace of a work.

A compliance with the first principle given for fortifications (Art. 7) would require that the work be placed upon the high ground, rather than upon the low; a compliance with the second principle would require that the parapets be placed near the brow of a height and not away from it. Hence, it is seen that the trace of a work generally follows the brow of the high ground forming its site, and is placed so that the work can sweep with its fires the slopes in front of the parapet.

155. Three cases will arise. The case in which the slopes of the site are gentle, easily ascended, and can be swept by front and direct fires of artillery. The case in which the slopes can be ascended by infantry without difficulty, but are too steep to be swept by a direct artillery fire. And the case where the slopes are so steep as to be ascended only by climbing.

The first case, or when the slopes are gentle, is the one most frequently under consideration.

156. Modification of profile.—Suppose a field work to occupy a surface which is level or slightly undulating considerably above the surrounding country, but joined to it by gentle slopes. It will be easily seen that a change of position in the lines of a work upon this plateau can only be made by modifying the profile.

Take a site like that shown in Fig. 73, and suppose the general directions in which a strong direct fire must be had (second principle of trace, Art. 36), to be indicated by the lines X Y, X_1 Y_1, and X_2 Y_2, (Fig. 73). The lines A D, A B, and B C, drawn perpendicular to the lines just named, would indicate the directions that the interior crests should have to get these direct fires.

The next point is the placing of these lines so that the fire from them shall sweep the slopes in their front.

Suppose a profile, X′ Y′ (Fig. 74) to be made representing the section cut out by the plane X Y. (Fig. 73.)

Fig. 73.

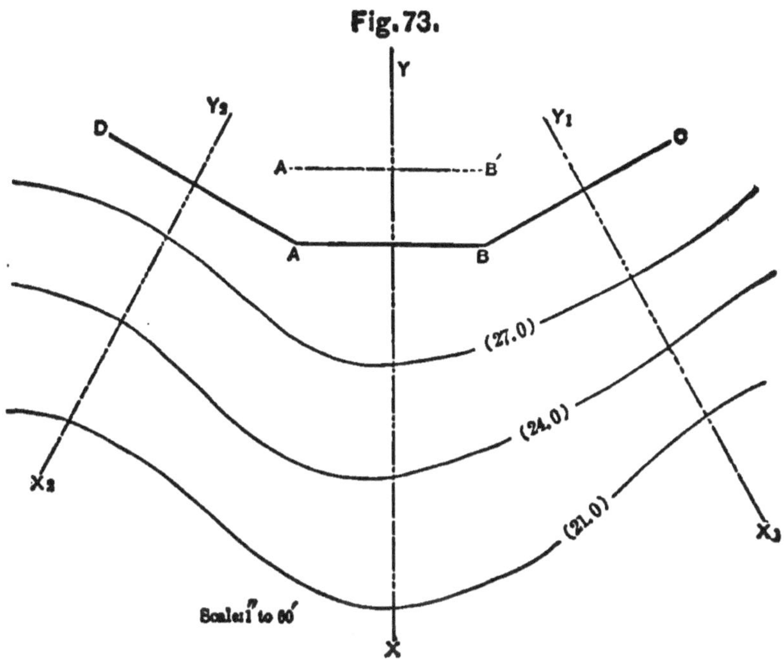

This profile gives the inclination of the slope con-

Fig. 74.

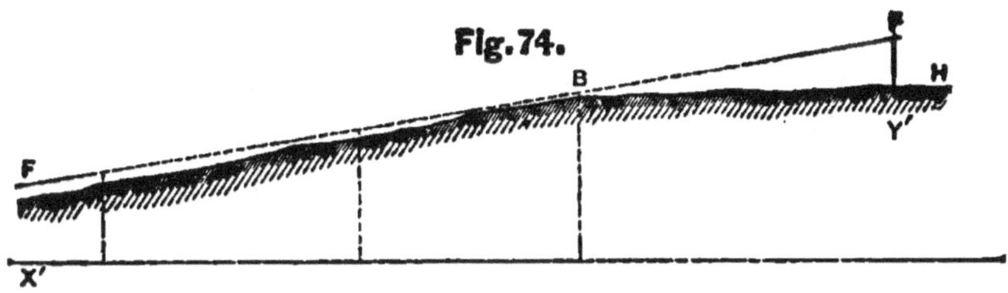

necting the lower level with the high ground. An examination of the profile makes it plain that a work placed on the plateau, B H, should not be too near

B, nor too far away from it. If too near, the guns can not be depressed enough to sweep the slope, B F. If too far, the brow, B, of the high ground will intercept the shots and make a dead space of the slope.

If a straight line, F B, be drawn tangent to the slope, or so that no point of it should be more than eighteen inches from the ground, and the interior crest of the work be placed upon it, it is plain that the whole slope can be swept by a direct fire of the work, so long as the angle made by this line with the horizontal is not greater than that made by the superior slope of the parapet.

If the face, A B (Fig. 73) occupies such a position that the interior crest, eight feet above the ground on which it is placed, is in this line, F P (Fig. 74), then it is evident that the slope can be thoroughly swept by its fire if the inclination of the slope is not greater than that of the superior slope.

It will also be seen, if the vertical at P (Fig. 74) represents this position of the interior crest, that moving P nearer to B can only be effected by lowering the parapet; or, moving it away towards H, by raising the parapet; otherwise the slope would not be swept by a direct fire from the face A B.

The foregoing sufficiently explains how the profile may be modified, with respect to the height of the parapet and inclination of the superior slope, for the purpose of obtaining a direct fire.

157. Modification of trace.—It is not always practicable, by moving the parapet, to arrange the trace of a work so that a direct fire can be obtained upon the slopes in its front. This is particularly the case in steep slopes, and upon ground in which the contours are very irregular in form. Slopes unseen by direct fire may be swept sometimes by a flanking fire from other parts of the work, or other works built for the purpose. Circumstances will decide as to the method which will best effect the object.

As an example, take the case of a piece of ground with an irregular outline, like that shown in Fig. 75.

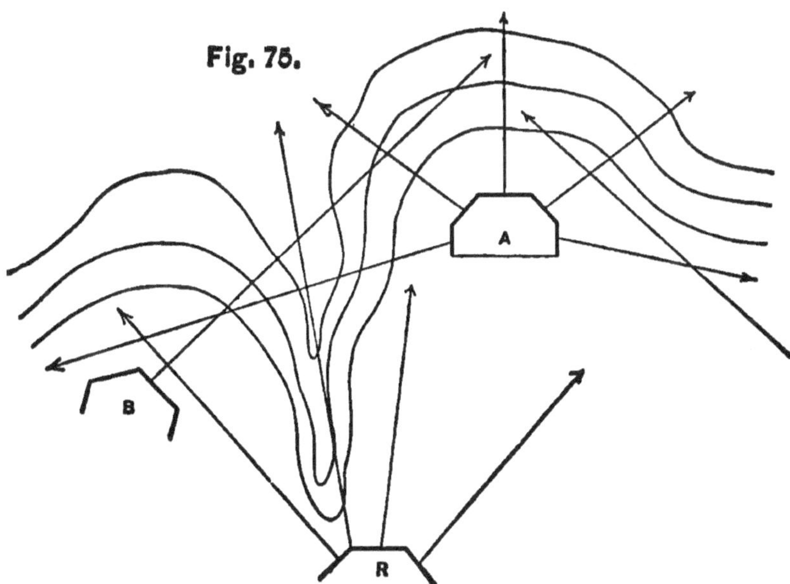

The plan of laying out the trace parallel to the brow would not be a good one, because of the great development of interior crest which would result, and even with that, a failure to obtain all the direct fire needed.

A better plan would be to modify the trace, and to make use of a line with intervals.

A line of works like B, A. etc., might be placed upon the salient spurs and in defensive relations with each other. A second line, as R, etc., might be used to defend the ground between the works in the first line, and to reach points not seen by the fire from these works.

Or, if it is desirable to use a "continued line," and it is not expedient nor judicious to follow the brow

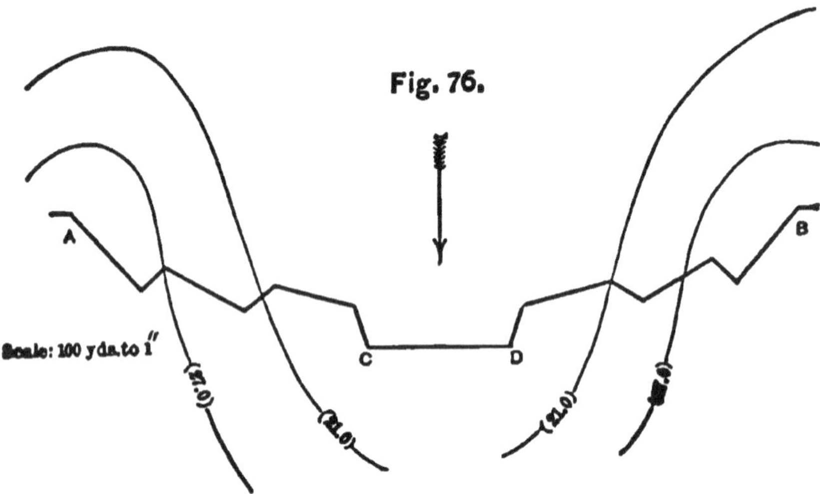

Fig. 76.

of the heights, a trace like that shown in Fig. 76 may be used. The general direction in which the fire of the enemy is to be expected is shown by the arrow.

The trace, instead of following the brow, runs from A to B along ground considerably lower than these points. A strong work should be built upon each of the points A and B, and an indented line receding, as it descends the slope, should join these works. The long branches should be directed so as to be safe from enfilading fire, and the short branches arranged to bring flanking fires upon the slopes in front of the faces.

The advantages of such a line when compared with a straight line joining the two points, A and B, are manifest.

158. Defence of steep slopes.—The brow of the height where the slopes are too steep for direct artillery fire may be broken, forming spurs; or the brow may be regular in form.

In the former case, works are placed upon the spurs, and the slopes in front of the works are swept by fires from the adjacent works, or by auxiliary works constructed for the purpose. In the latter case, the works should be arranged to bring as strong an artillery fire as practicable upon the approaches to the foot of the hill, and supplemental flanking arrangements should be used to sweep the slopes with musketry fire.

Under circumstances like these, the works are not usually exposed to a close artillery fire, and they are not therefore required to be so strong. The profiles may be modified accordingly.

159. Defence of precipitous slopes.—When the slopes are so steep as to be ascended only by climbing by using the hands in the ascent, their defence is an easy matter.

The works to be constructed, unless exposed to cannon fire, may have a very slight profile. A parapet four feet and six inch inches high, and thick enough to resist musketry, will be, as a rule, all that is necessary.

Ditches, as obstacles, will not be necessary. A sufficient obstruction can be obtained by **scarping** the slope, effected by cutting away its face until a steep slope is made. If a ditch is used, it will be placed ordinarily in rear of the parapet.

If the high ground terminates in a plateau or large flat surface, it would be well to establish a line of strong field works, some two or three hundred yards in rear of the brow, which would command the works placed along the crest of the slope.

160. General plan.—The defence of a position is made in accordance with some general plan, which plan decides as to the predominance of the offensive or defensive features of the position, the points to be fortified, the number of troops that can be spared to occupy the fortifications, and the time in which the works must be built.

After these general points are fixed, the engineer may then consider the kind of works which will be best fitted for the purposes intended, and arrange the profile, defilade, and construction according to the natural features of the site and the means at his disposal to build the works. The position should be well selected, since no amount of skill can remedy defects which are fatal. The faults may be ameliorated, but the position can not be made a strong one if inherently weak. The engineer shows his skill in adapting his constructions to those positions which in themselves have some merit. This adaptation requires a thorough knowledge of the principles of fortification and of the details of the art.

How extensive these modifications may be can be seen by the examples just given.

Bridge-heads.

161. Definition.—The term, **bridge-head,** (French, *tête-de-pont*) is applied to any field work or line of works which is built to defend a crossing of a river at a particular point, and to prevent its use by an enemy.

These crossings are made by means of bridges, ferries, or fords. Of these, the bridges are the most reliable for all stages of water, are the most convenient, and are the constructions which are usually guarded and defended with the greatest care. The principles governing their defence are equally applicable to the defence of fords or ferries.

162. Object.—The object to be attained by the use of a bridge-head is to protect the bridge from the destructive measures of an enemy, and to prevent its use by him.

This preservation of the bridge may be for the purpose of keeping open the communication with the opposite bank, with the expectation of using it either by small parties or by large bodies of troops.

When intended for the use only of small bodies of troops, and to keep possession of the line of communication, the bridge-head used for the defence of the bridge, may be a simple field work, generally a redan, or a lunette. (Fig. 77.)

Fig. 77.

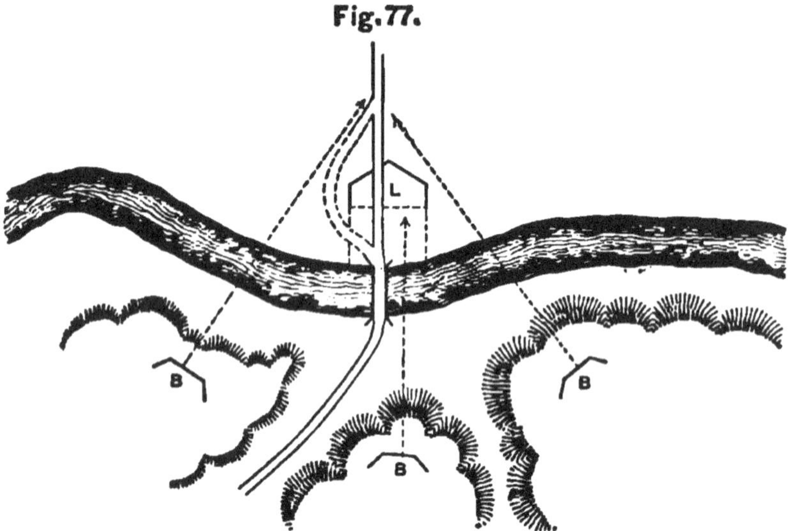

This half-closed work, a redan or a lunette, L, is usually placed on the road leading to the bridge and upon the side of the river towards the enemy.

Its rear should be closed by a stockade, and this stockade connected with the approaches to the bridge, or banks of the stream, by a palisade. A temporary road, as shown by the dotted lines, should be built to allow of a free passage around the bridge-head, instead of going through it. If there is no danger of the enemy appearing in force and with artillery, a block-house may be built within the bridge-head.

The faces and flanks of the bridge-head should be placed so as to intercept all fire that the enemy can bring upon the bridge from any position which he might occupy.

If there is danger of the enemy coming in force, either to destroy the bridge or to cross the river at this point, the bridge-head should be made strong enough to resist an artillery fire, and should be strengthened by batteries placed in positions like B, B B, by means of which a concentrated fire can be had upon the ground in front of the bridge-head, and a fire into the interior of the work, as indicated upon the figure.

Bridge-heads of this kind should be made strong enough to hold the enemy in check until re-enforcements could arrive, or until the bridge could be made useless to the enemy.

Bridge-heads of this simple kind are useful to protect bridges which are or might be useful for reconnoitering parties composed of small numbers.

163. Strong bridge-heads.—A bridge-head used to guard a crossing when the army may move in force, either to make an advance movement, or to retreat, must be made **strong**, if the enemy is, or is expected to be, within its immediate neighborhood.

The presence of bridges already constructed may determine the selection of the point at which a general may desire to cross his army, either in making an advance towards the enemy, or retreating from him.

If he has to have bridges built, he will have some latitude in the selection of the point at which he may cross.

The selection of this point will be governed by strategical, tactical, and technical considerations.

The tactical requirements will be best satisfied, as a rule, when the point of crossing lies in a bend, with its convexity towards the army which is to cross.

The bend will in many cases allow the establishment of batteries from which flanking and cross fires can be had upon the ground in front of the bridge head (Fig. 77).

The banks in a bend are usually unequal in height, the higher bank being on the convex side and opposite to the bridge-head. This difference of height will frequently allow the site of the bridge-head to be commanded by the fire from the opposite bank.

There are other prominent tactical advantages accompanying a bend, such as the approaches to the river on the convex bank being concealed from view of those on the opposite side; a greater protection being obtained for the bridge structure; etc.

Great as these advantages are, a good crossing requires in addition, if it is to be used by troops moving in force, a roomy space in front of the bridge-head hidden from the view and sheltered from the fire of the enemy in which the masses can easily deploy. (Fig. 78.)

Fig. 78.

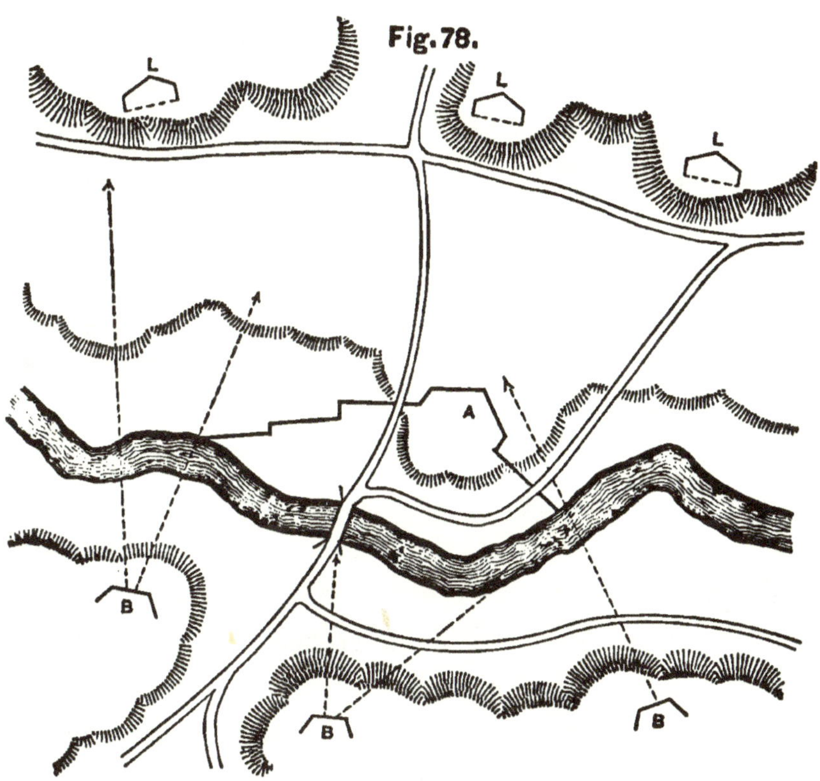

In this example, the shelter is given by a line with intervals, L, L, L, placed some distance in front of the bridge-head and encircling the space to be occupied by the troops in their movements forward from this point, or in retreat. The line should be constructed in accordance with the principles already laid down; the works, in defensive relation with each other, and, where necessary supplemented by secondary works.

The extremities of the line should, as in this case, rest upon the river or upon points which prevent its being turned.

A bend in the river enclosing a narrow strip of land would be exactly fitted for a bridge, and for its bridge-head, but would not be so favorable for the passage of large bodies of troops, because of the limited space afforded to the masses for deployment after having crossed the river.

There is shown only one bridge in the figure. In case of a rapid advance being made, or a quick retreat closely followed by the enemy, several bridges should be used.

164. Horn works, etc.—The traces used for bridge-heads are of various kinds, and are modified by the irregularities of the site, by the directions of the approaches, and by the forms of the banks. Redans, lunettes, bastioned lines, cremaillère lines, etc, of various shapes and forms, are used.

Certain arrangements of the bastioned lines give rise to the bridge-heads known as **horn works,** and **crown works.**

Let a point be taken in front of the bridge and some distance from it. Through this point let a right line be drawn perpendicular to the general direction of the bridge. On this line thus drawn as an exterior side, let a bastioned front be constructed, and its salients be joined with the banks of the river by straight lines, which are so directed that they can be swept by a fire from the opposite bank. The resulting trace is known as a horn-work.

It is plain that this work will only be used when the main approach to the bridge is in the prolongation of its length.

If through a point assumed in front of the bridge two right lines were drawn making an angle with each other, and prolonged until they reached the banks of the river, and on these two lines, as exterior sides, oastioned fronts were constructed, the resulting trace would be that of a crown-work.

It is plain that this trace will be employed when the approaches to the bridge are oblique to the direction of the bridge, and that the enemy would use one as quickly as the other.

If there are several approaches, and the entire front is exposed to attack, a continued bastioned line might be used, enclosing the space in front of the bridge from bank to bank. In this case, if a salient occupies the central position, the line is known as a crown-work. In the latter case it is called a

complex crown-work, to distinguish it from one constructed on two sides only, which is called a **simple** crown-work.

When the bridge crosses the river at a point where there is no bend, it is frequently the case that works are constructed at both extremities of the bridge. When this is the case, the works form what is known as a **double** bridge-head, to distinguish it from those just described, which are termed **single** bridge-heads.

CHAPTER XIV.

HASTY INTRENCHMENTS.

165. Hasty defences.—Hasty defences include all extemporized shelters, which are quickly constructed (in a few hours at most) from materials found upon the spot where the shelter is needed.

In consequence of the effectiveness of modern fire-arms, a body of troops can not retain a close formation for a single hour even, if in the presence, and exposed to the fire of an enemy in force. The men are forced to seek shelter, by lying down on the ground, or by crouching behind any slight inequality which may exist in the surface, or behind some kind of a screen which they may be able to construct. The screen may be two or three logs rolled together (Frontispiece), a heap of fence rails, a slight mound of earth, or anything whatever its nature which will hide the soldier from the enemy's view.

166. Shelter-trenches.—The simplest form of shelter, for a soldier in open country, under ordinary circumstances, is a shallow trench, which will furnish from the excavation, sufficient earth when heaped upon the side towards the enemy, to screen the soldier in the trench from the enemy's view. Trenches of this kind are known as **shelter-trenches**, but are most frequently called by the soldier, **rifle-pits**. Because of the shortness of time required to build these defences, they are known under the general name of **hasty intrenchments**.

Hasty intrenchments were much used by both of the contending armies in the late war in the United States. They were used so frequently, and found so efficacious, that the men acquired the habit of intrenching their line immediately upon halting after a day's march, if the enemy was near. No compulsion, no orders, even, were necessary for the men to begin this work; the main difficulty was to make them delay long enough to allow a proper trace to be marked, by which they might be guided in the construction of their line. Instances are known, where the men, not having intrenching tools, executed the trench with the bayonet and tin cup. These shelter-trenches, thus rudely constructed, were deepened and strengthened until they were able to resist field artillery, if the position was to be occupied for any length of time.

Slight as these defences were during the early stages of their construction, they formed, when defended by good troops, an obstacle difficult to overcome; and they were captured only by extraordinary effort, accompanied by a great loss of life on the part of the attacking forces.

167. Construction.—The trench represented in

Fig. 79 is the smallest that can be made, upon level ground, and afford shelter to a man.

Its depth in rear is one foot; its width is five feet; and its length is dependent upon the number of men using it. The earth taken from the trench and heaped upon the ground in front will make a mound fifteen inches high and at least two feet thick, affording a screen from the enemy's view, and a tolerable shelter against musketry fire.

A log laid in front of the trench and the earth thrown over and against it, adds materially to the protection afforded by the shelter.

A trench of this kind can be executed by the soldier with a shovel in thirty minutes.

168. In one hour's time, a soldier can deepen and strengthen this trench, giving to it a depth of eighteen inches throughout its width of five feet, and raising the mound to a height of eighteen inches. (Fig. 80). This additional depth of trench and height of mound allow the soldier using it, to take a kneeling position when firing; a position more convenient than the reclining one necessary in the last case.

169. In from two to three hours' time, if the soil is not difficult to dig, the trench can be enlarged to a width of eight feet, and the earth excavated thrown upon the mound, raising it to a height of three feet with a thickness of about four feet. (Fig. 81). A trench of these dimensions allows the soldier to occupy a standing position when firing and it approximates to the form of trench to be constructed finally, if there is time for the purpose, and the position is to be held.

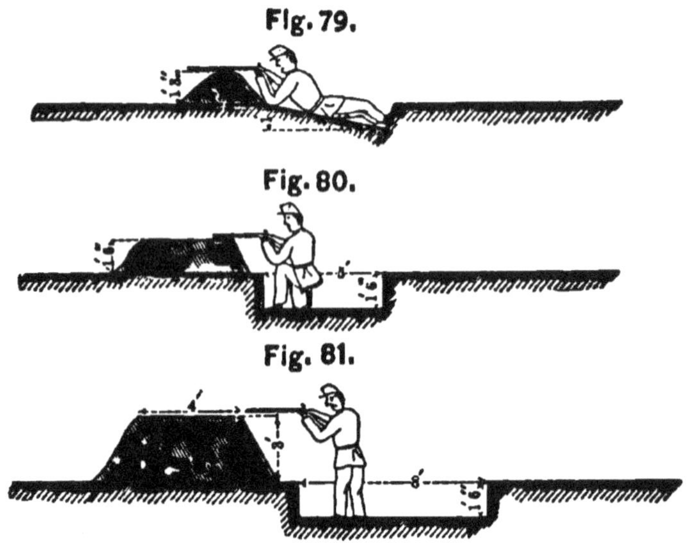
Fig. 79.

Fig. 80.

Fig. 81.

170. It will be observed that the first trench offers no serious obstruction to the advance either of artillery or cavalry. In the latter forms, the trenches will obstruct, more or less, forward movements, if the trenches form a continuous line.

At short distances. intervals should be left in the line of the trench, or ramps be arranged in the trench, so as to allow the artillery, etc., to march straight over the intrenched line if a forward movement is to be made.

171. Covered Communications.—Shelter

trenches are much used to afford covered communications, along a given front; to connect the works in a "line with intervals;" to bring a musketry fire upon ground which can not be swept by the fire from a particular work; etc.

The least dimensions of a shelter trench, when used as a communication for infantry only, are given in Fig. 82.

Fig. 82.

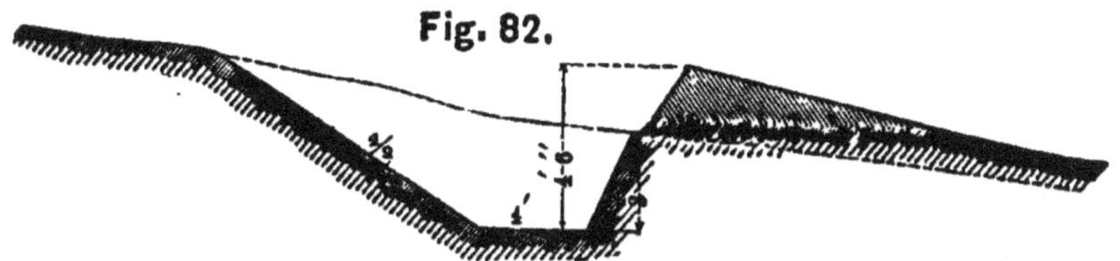

The trench should be made three feet deep, and four feet wide at the bottom. The earth should be thrown on the side towards the enemy, and then levelled off in the form of the superior slope of a parapet, so that the men in the trench can fire over this mass of earth.

If the trench is to be used for the passage of artillery, or to be used by bodies of troops passing from

Fig. 83.

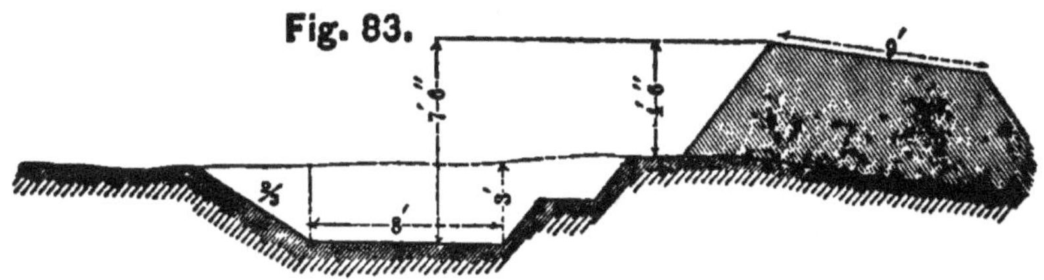

one point to another along the front, its dimensions should be those shown in Fig. 83.

The least width at bottom should be made eight feet, and the height of the top of the mound of earth should be at least six feet and a half above the bottom of the trench. The side of the trench toward the enemy should be cut into off-sets, and arranged so as to allow a fire of musketry over the parapet.

172. Trace of shelter trenches.—Shelter trenches are rarely made to follow a straight line, but usually conform to the contour of the ground. The trace should be marked on the ground if there is time to do it. It will economize the labor of the troops, and avoid an unneccessary waste of time.

The trace should be governed by the general rules laid down for field works, and great care should be taken that it can not be enfiladed by a fire of the enemy.

Profiles are not necessary. The points which would be occupied by them may be marked by men standing upon the edge of the proposed trench towards the enemy. A line would then be marked on the ground, by a pick, passing through the points selected. Parallel to this line and twelve or fifteen feet in rear of it, the line of troops should be formed. The front rank, furnished with intrenching tools, would begin the digging; the rear rank would lie down. Reliefs should be formed, and the trench rapidly executed.

173. Shelters for artillery.—The shelters for artillery or cavalry may be made in a very short time, in a way similar to that shown for the shelter trench for infantry.

On undulating ground, the shelter trench for infantry is frequently on the slope; the shelter for artillery would generally be on, or behind, the crest. (Figs. 84 and 85).

Fig. 84.

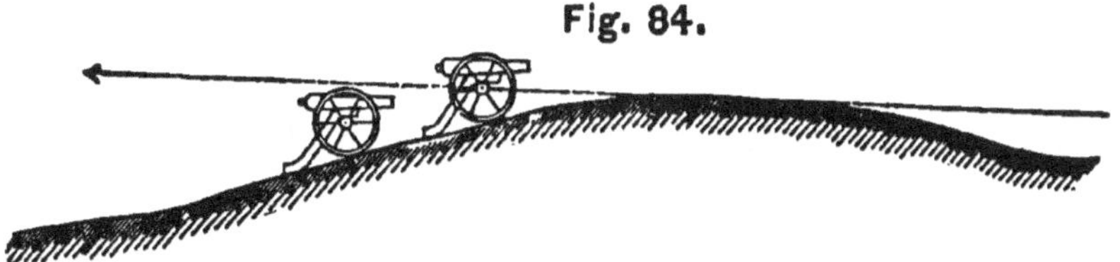

It is seen that a piece of artillery on the crest of undulating ground can be quickly run under cover, if it be desirable. (Fig. 84).

This cover can be quickly and easily improved, by making a slight excavation and arranging a mass of earth in front of the gun. (Fig. 85).

Fig. 85.

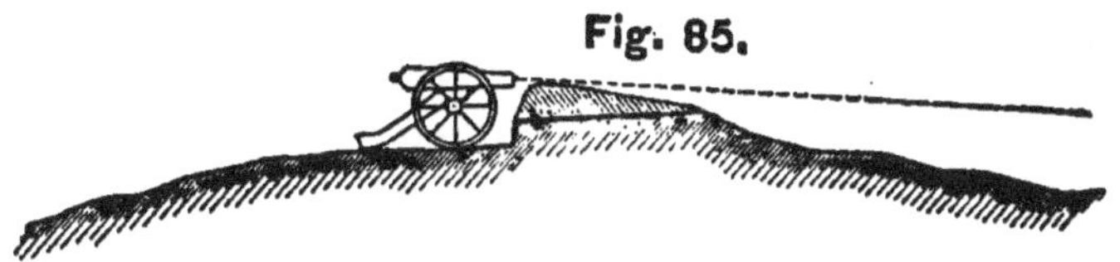

Slopes of this kind could be used for infantry as well as for artillery; and where a simple screen is the main object to be had, the communication would be along the reverse slope.

174. Defence of a house.—It frequently happens that the front of a defensive position is occupied by one or more houses. The houses may be on the line of battle, or they may be close to it.

If they can be readily put in a state of defence, they are so arranged ; if not, they should be torn down.

A house, solidly built of stone or brick, and which will not be exposed to artillery fire, can be readily made into a defence of considerable strength. Brick houses with slate roofs are the best for the purpose.

The first thing to be attended to, in putting a house into a defensive condition, is to clear the space in front of it of every thing which would screen an enemy's approach. The next is to loop-hole the walls, making at least two tiers of fire in the lower story. The loop-holes in the upper stories should be arranged to get a fire as close as possible to the foot of the walls of the building.

Ditches cut near the walls, and the earth thrown against them, are recommended to keep the enemy off and to add to the resistance of the walls.

All doors, sashes of windows, etc., inflammable in their nature, should as far as possible be removed and the openings barricaded by sand-bags, or by boxes and barrels filled with earth.

Barrels filled with water should be placed in each room to put out any fire which might happen to break out in them. Earth in a moist state might be spread upon the floors, or on any of the flat surfaces, liable to take fire, to prevent their burning.

175. Tambours.—A house which has projecting parts, like bay windows, can be readily arranged to give a cross fire in front of its salient angles. The same result may be obtained where two houses flank each other. Where flank defence can not be thus obtained, it may be had by using constructions known as **tambours.** (Figs. 86 and 87.)

Fig. 86.

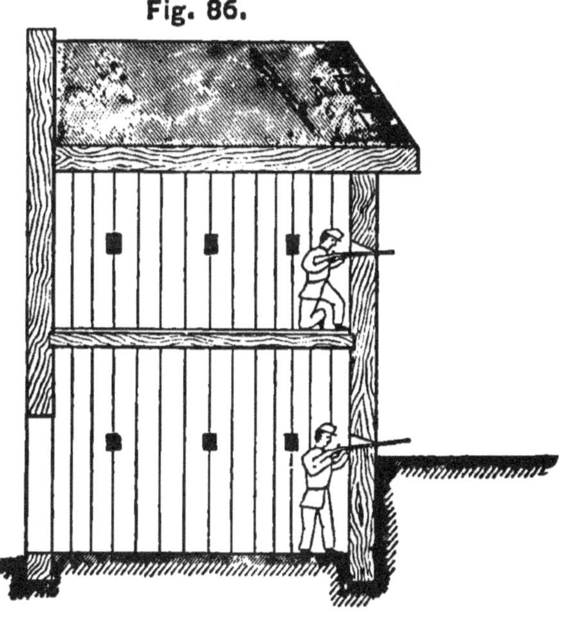

A tambour is a stockade, generally redan-shaped in plan, placed in front of a door or other opening, in the wall to be flanked, and arranged with loop-holes to bring a fire upon the ground in front of the salients of the line to be defended.

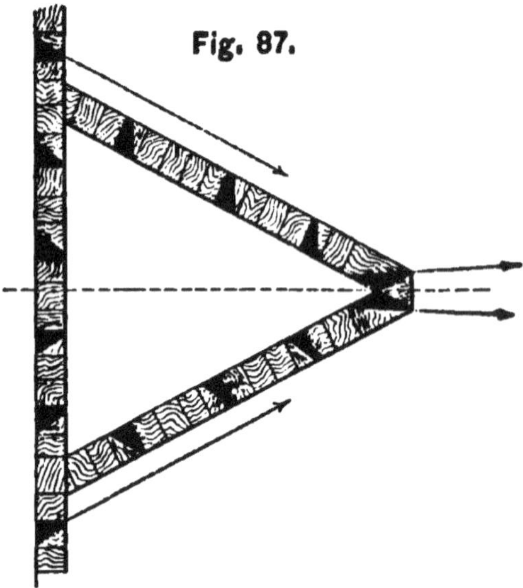
Fig. 87.

It may have one or more tiers of fire; it may be one story high, or it may be two.

176. Machicoulis galleries.—Where tambours can not be employed, a flank defence may be obtained by means of a construction known as the **machicoulis gallery**. (Fig. 88). The object of this construction is to bring a fire upon the ground along the foot of a wall.

Where balconies exist, galleries of this kind can be made from them.

If no balcony belongs to the house, the gallery may be formed by breaking two or more holes through the wall, on a level with the floor of the second story.

Through the holes thus made, stout beams are passed and their inner ends firmly secured to the joists of the floor. These beams form the joists of the gallery,

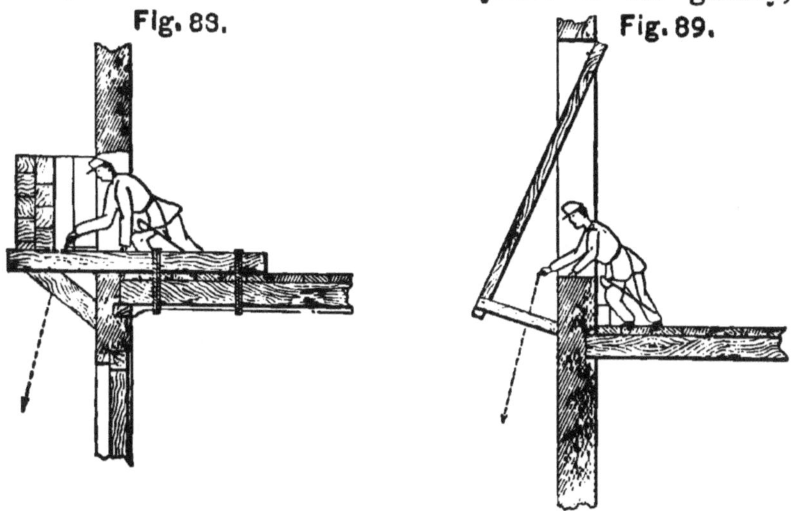
Fig. 88. Fig. 89.

and they may be braced, or not, by diagonal pieces as shown in Fig. 88. On the outer extremities of these joists, a musket-proof shelter is constructed; this shelter may be made of two thicknesses of plank, enclosing some light resisting material.

The men fire through the bottom of the galleries upon the ground beneath, using pistols, if they have them, in preference to the musket or the rifle.

177. Expedients similar in principle to the machicoulis gallery are used to bring a fire upon the ground in front of the wall of a house where the gallery can not be used. One of these is shown in Fig. 89, where the sash of the window is removed and a musket proof shelter is inserted in the window at an angle so as to leave an opening at the bottom through which a man can fire his pistol.

178. Stone walls, hedges, etc.—Obstructions like stone walls, hedges, etc, can be utilized in the defence of a position.

High stone walls can be loop-holed, and arranged for defence in a way similar to that adopted in stockades. Low stone walls may have shallow trenches dug behind them, and the earth thrown over and against the wall. The top of the wall can then be crenelated or arranged with improvised loop-holes.

Hedges can be made into serious obstacles to an enemy's progress, and can be quickly converted, by means of earth thrown against them, into good shelters, for the defence.

179. Woods.—If the woods are too far to the front to allow of their being defended, they should be cut down, or "slashed," thus forming an obstacle, but not a screen to the enemy.

If they are to be defended, a line of shelter trenches may be constructed along the outer edge of the woods, on the side towards the enemy, but within the edge so as to be concealed from his view. A thin strip of the timber may be cut down to form an abatis in front of the intrenchments.

Good roads and open spaces should be arranged in

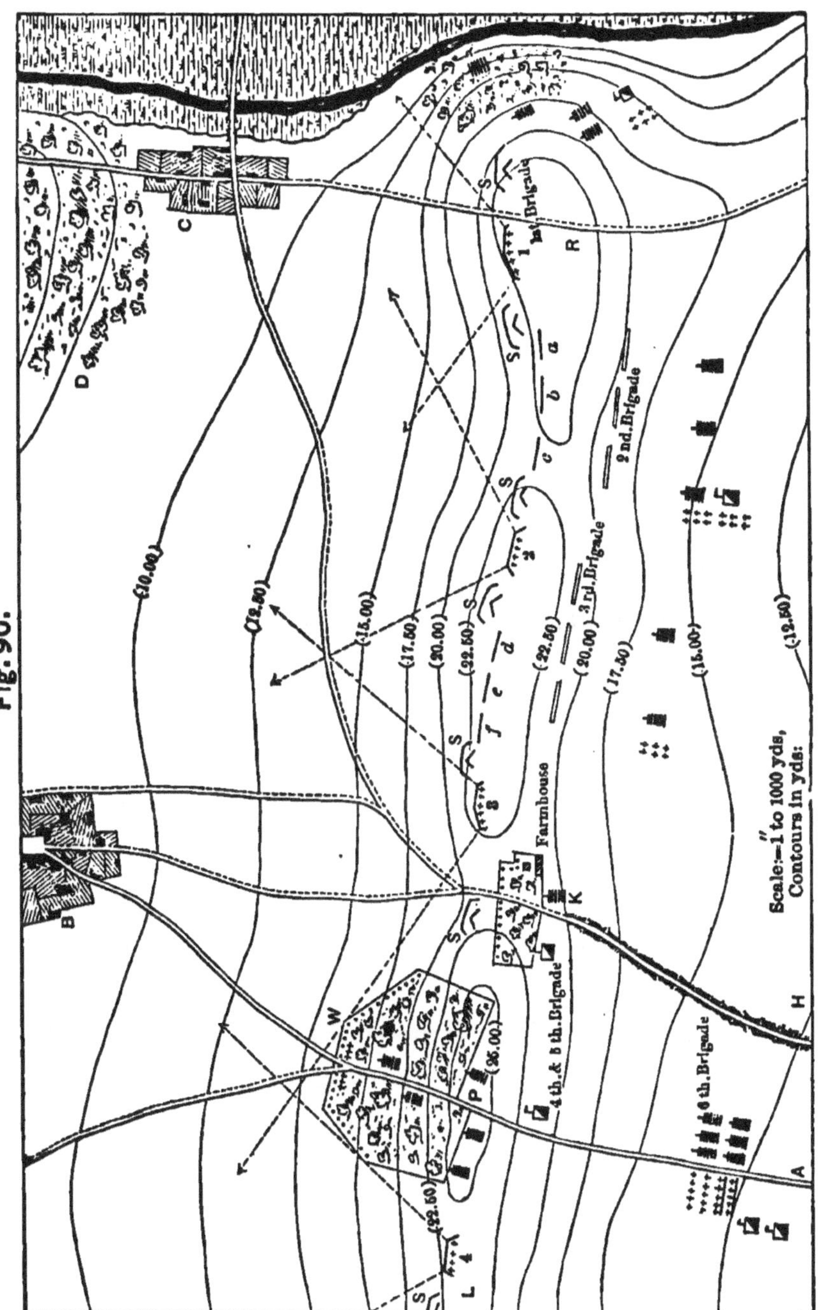
1st. Brigade
2nd. Brigade
3rd. Brigade
4 th. & 5 th. Brigade
6 th. Brigade
Farmhouse
Scale:—1 to 1000 yds,
Contours in yds:
(10.00)
(12.50)
(15.00)
(17.50)
(20.00)
(22.50)
(25.00)

Fig. 90.

the woods to allow the troops and guns to move about freely.

180. Roads.—The roads in rear of a position should be arranged so that they can be freely used by the defence. Those in front, should be obstructed, and in every way made useless to the enemy. Bridges to be used by the defence should be put in good order; those which would be of service to the enemy, should be destroyed.

181. Intrenched fields of battle.—All hasty intrenchments and defences have their primary use in defending a position which is to become, in a short space of time, the scene of battle.

The principles already named, as necessary to be observed in fortifying a position for defence, are equally applicable to the methods used to intrench a line on the field of battle. The more completely the general conditions for fortifications are fulfilled, the stronger will be the intrenched line.

The example shown in Fig. 90 may be taken as an illustration of what is meant by an intrenched line of battle.

An army corps is supposed to be marching on the road from A to B (Fig. 90), and the commander is instructed, if the enemy appears in force, to take a position and hold him in check.

The position selected may be one which is to be held only for a short period, so as to allow time for the army to concentrate, or to force the enemy to concentrate his forces to attack; or it may be one in which a decisive battle is to be fought.

It is supposed that the commander of the army corps, when in the vicinity of A, learns of the approach towards B of the enemy in force, and he feels obliged to hurriedly occupy a position to resist the enemy's advance.

A reconnoissance has shown him, that the best natural position for him to occupy for this purpose, is the ground between L and R, and that the distance between these points is about four thousand yards.

He finds that the high ground runs nearly perpendicular to his line of retreat through A, and is bounded upon the right by an unfordable stream with low and marshy banks. He finds the ground higher where the main road from A to B crosses it, than at any other point and that it slopes gently both to the front and rear, commanding all ground within cannon range.

The natural features of the ground are shown in the sketch.—A wood W of considerable size in front of the centre of the left of the position; a village at B, outside of cannon range; a wood at D, also outside of cannon range; a small collection of buildings at the cross-roads at C; a wooded slope upon the right of the position; a farm-house, capable of being put in a defensive condition, almost on the general line to be occupied; etc, form the most prominent features of the ground.

Good roads traverse the country; the one leading from A to B, being a turnpike; the others, common roads such as are seen generally in the country. The road from H to K has worn into the surface, forming a sunken, or what is generally known as a **hollow** road. The country is open and cultivated; intersected with ordinary farm fences, and divided into fields for pasturage, cultivation, etc.

The commander ascertains from the reconnoissance that the high ground at P must be held at all hazards; that this part of the line forms the key point of the position. That the woods at W are too large to be "slashed" in the time at his disposal. They must therefore be defended and made so strong as to remove all danger of the enemy's obtaining possession of them.

The number of troops forming the army corps is about thirty-four thousand, and is sub-divided into divisions, brigades, etc. For simplicity of details, the corps will be supposed, in this example, to be divided into six brigades, of six regiments each, and its com position to be as follows:

Six brigades of infantry, three regiments of cavalry, twelve batteries of field artillery, three batteries of horse-artillery, and six companies of engineer troops. These will number about as follows:

Thirty-six regiments of infantry, ...	27,000 men
Three regiments of cavalry, ...	3,600 "
Fifteen batteries of artillery, ...	2,250 "
Six companies of engineer troops, . . .	900 "
Total.	33,750 "

Since the left of the line is to be the most strongly defended, the commander distributes his infantry as follows: one brigade upon the right; two brigades upon the left; two brigades, to defend the ground between the right and left, and the remaining brigade in reserve.

The brigades are placed about as follows: The first brigade being expected to defend the high ground at R, may be posted as follows: two regiments in the intrenchments *s*, *s*, on the right and left of the battery at 1; one regiment in the skirt of the woods on the right, supported by another regiment near it but hidden by the slope of the hill; two regiments should be placed a short distance in rear of the right, to form a second line and act as supports to the first line. The second brigade is placed on the left of the first, as follows: three of its regiments deployed behind the crest of the hill; the remaining three, the troops being in column, should be in rear of the others, forming a second line. The third brigade is placed on the left of the second, as follows: three of its regiments deployed along the slope upon the same general line formed by the second brigade; one regiment placed in the works *s*, *s*, near the battery at 2, as a support to the guns; the remaining two regiments in rear, forming a second line. The fourth and fifth brigades are to defend the left. The whole of the fourth brigade is posted in the woods **W**, three of its regiments deployed along the skirts of the woods on the sides towards the enemy, and the remaining three, in column, in convenient supporting distances. The fifth brigade has one regiment posted at the farmhouse; one, in the intrenchments *s*, *s*, supporting the batteries at 3; one, supporting the battery at 4; and the remaining three regiments behind the woods, as shown in the figure. The sixth brigade forms a reserve, and occupies a convenient position in rear near the road coming from **A**.

The artillery is posted as follows: two batteries upon the high ground on the right, commanding the approaches in that direction,—marked on the figure at 1; one battery, in the interval between the second and third brigades, at 2; two batteries, on the left of the third brigade at 3; one battery on the left at 4; one battery, in the front edge of the wood **W**; two batteries, in the second line behind the second brigade; one battery in the second line behind the third brigade; one battery of horse artillery with the cavalry on the right; and the remaining batteries, four in all, with the reserve.

The cavalry is posted as follows: one regiment behind the right wing; one behind the left wing; and one with the reserve. The cavalry on the right may be divided into two parts, one part is posted behind the extreme right, the other behind the second brigade. The cavalry behind the left wing may be divided into two parts, one is posted behind and near the farmhouse, the other behind the wood **W** and near the main road.

The engineer troops are distributed along the entire front, engaged in the preliminaries belonging to the work of intrenching the position. These preliminary operations consist in putting the woods **W** in front, the woods upon the right, and the farm-house, in a condition of defence; in laying out the shelter trenches, *a*, *b*, *c*, etc., and posting the working parties detailed to construct them; and in tracing the epaulments, 1, 2, 3 and 4, and the field works, *s*, *s*, *s*, etc., intended for the batteries and their supports.

The woods **W** are put in a state of defence as described in article 179. The battery posted in the edge is placed so as to sweep with its fire both of the roads leading to the woods. The batteries at 3 and 4 are placed so as to cross their fire upon the ground in front of the woods.

The woods on the right are arranged in a similar manner to afford an obstacle to the enemy moving in that direction.

All obstructions existing upon the ground in front of the position which can be made to interfere with the movements of the enemy are preserved and added to; all those which would screen his movements or afford him shelter, are removed or torn down.

The hollow road in rear should have its banks cut down so as not to interfere with the movements of the troops in crossing from one side to the other.

The farm-house is arranged as described in article 174. The shelter trenches and the epaulments for the batteries are constructed as previously described.

The field works, *s*, *s*, *s*, etc., have the traces of blunted redans, or of lunettes with obtuse salients. The faces should be made from seventy to eighty yards long, and so directed as to sweep the slopes in front. It is recommended to place in each work, a redan-shaped, defensive traverse arranged as indicated in the figure, and to give it a command equal to or greater than the parapet of the work.

These works should be placed, so as to leave an interval of not less than one hundred yards, between their flanks and the batteries they support.

The profile of these works, *s*, *s*, etc., should approximate as nearly as possible to the normal profile. The time disposable for their construction will decide upon the command they shall have. By placing working parties upon both sides of the parapet, constructing the part in rear of the interior crest with the profile shown in Figs. 81 and 82, and constructing the part in front with the normal profile, a strong work can be quickly built.

It will be seen, that the commanding general, by the use of these intrenchments thus hastily constructed, can concentrate, without risk, the greater part of his corps upon the left, the key point of his position. By their use, he is able to supply the want of numbers and thus strengthen his line. The principles observed in this example apply equally to a position of greater or less extent.

CHAPTER XV.

ATTACK AND DEFENCE OF FIELD FORTIFICATIONS.

182. Attack.—An attack made to capture a field work, or to carry a line, may be a quick, sudden effort, or it may be a prolonged one. The former is known as an **assault**; the latter, a **siege.**

Assaults are of two kinds; open assaults and those made with great secrecy.

Whatever be the kind of assault, it should be preceded by reconnoissances, made as full as possible, for the purpose of ascertaining the best and easiest approaches to the work, the nature and position of the obstacles, the numbers and kinds of troops composing the garrison, and the strength and positions of the reserves exterior to the work, but near enough to take part in its defence.

Particular attention should also be paid to the positions for the artillery of the attack. These positions should be such that the guns can bring enfilading fires, on the principal faces of the work, strong cross fires upon the point of attack, and if possible, a sweeping fire on the approaches to the work in rear. An important point, to be observed in this matter, is to select, if possible, positions from which the guns will not have to be removed during the attack.

183. Open assault.—An assault of this class is made suddenly, and if possible, without the enemy's knowing it until the rush is made. It is an open assault, because the approach is made openly, though the attack be unexpected.

The assault may be divided into three periods, as follows :

1. The preliminary operations, and the artillery attack.

2. The advance of the attacking troops from the cessation of the cannonade, until they arrive at the counterscarp of the ditch.

3. The assault of the parapet, and the capture of the work.

184. The preliminary operations consist in selecting the points of attack ; designating the troops to make the attack, and placing them in position ; organizing bodies of pioneers, or engineer troops to remove obstacles; etc. It is supposed that the outposts of the enemy are all driven in, and his troops are within the work, or under its immediate protection.

A heavy, converging fire of artillery, within accurate range, has been deemed by military authorities to be an essential element of success in an open assault. The manuals all prescribe, in an open assault, an artillery attack preceding the forward movement of the attacking forces.

The objects expected to be attained, by this artillery fire, are as follows:

1. To silence the artillery of the defence.

2. To damage, and possibly to destroy the works.

3. To inflict losses, and to harass the men to such an extent, that they may become demoralized.

The first object is attainable. The theory of an open assault supposes a superiority of numbers upon the part of the attacking forces, and a superiority of artillery. When this is the case, the silencing of the artillery of the defence becomes a mere matter of time.

The second object is hardly practicable so far as its destruction is concerned, if the work is made of earth, and has a good profile. The fire may, however, be successful in tearing up the superior slope, destroying the improvised loop-holes, and producing irregularities in the interior crest of the parapet. The amount of damage which can be made, will depend upon the amount of fire that can be concentrated upon the work. The effect of the damage will be to impair, more or less, the confidence of the men, in the efficacy of the work to shelter them.

The third object may, or may not be, attained. The kind of shelters, the strength of the parapet, and the character of the troops are factors of the problem.

The conclusions drawn from recent experiences

are, that an attack by artillery, preliminary to an open assault, may be useful against works of weak profiles not provided with bomb-proofs, but against strong works with good interior arrangements, the artillery attack is of little service, and may be injurious to the attack so far as it serves the defence as a warning of the proposed assault.

185. The features of the second period are the advance of the attacking forces, and their progress to the ditch. The advance is made by a line of infantry, in extended order, preceded by skirmishers. A second, and even a third line, follows the first.

The skirmishers advance by "rushes," taking advantage of every inequality in the surface to shelter themselves, and being reinforced continually from the line in their rear. Accompanying the line of supports are pioneers, or engineer troops, provided with tools, etc, to remove or to make passages through all obstacles in the way of the approach. The skirmishers keep up a sharp fire upon the parapet, while the obstacles are being removed, and while the supports are forming in the ditch, or on the berm.

The methods of removing these obstacles are described in the books, and manuals, relating to practical engineering operations.

186. The third period includes the assault upon the parapet and all succeeding operations.

As soon as a sufficient number of the troops have

assembled in the ditch, or upon the berm of the work, the assault upon the parapet is made by a rush of these troops in a body up the exterior slope and over the the parapet into the work.

If a defence is still continued, a hand-to-hand conflict follows. This should result in a capture of the work in consequence of the superiority of numbers on the part of the assailants.

If the defenders should have left the parapet upon the assailants reaching the ditch, it may be that they have done so for the purpose of occupying some interior retrenchment, or block-house. In this case, the guns of the work, if any, should be turned against the retreating forces and steps be taken to attack the inner defences. An examination of the powder magazines should be made immediately, to guard against explosions either accidental or intentional.

Steps should be taken, as soon as the work is captured, to put it in a state of defence against recapture, if it is to be occupied, or to destroy it, if it is to be abandoned.

187. An attack made secretly.—The preliminaries of a secret attack are identical with those of an open assault, except in the secrecy of the movements which are made. The troops, who are to make the attack, are kept in ignorance of the object of their movements until they are assembled at the point from which they are to make the attack.

The success of the attack depends upon its being **unexpected** by the defence, and upon finding the defenders, in a measure, unprepared to resist it. An attack of this kind is a surprise, and is known frequently as an **attack by surprise.**

188. Success in either case, an attack made openly or secretly, is greatly dependent upon finding the enemy unprepared to resist the assault. The attack should therefore be made suddenly, and without the enemy suspecting the intention.

189. Attack by artillery only.—Under the circumstance that the attacking forces are overwhelmingly superior in artillery, it is possible, in works hastily constructed and not properly provided with interior defences, to force a garrison to capitulate by use of artillery alone, supplemented by sharp-shooters who may be able to get near enough to the works to pick off the gunners of the work.

190. Key-point.—Great care must be taken to direct the assault upon the **key-point** of the intrenched line or position. The mistake has been frequently make of directing the main attack upon the wrong point, which attack when successful, produced no lasting benefit to the victorious troops.

191. Defence.—Fortifications are inert masses, passive in their nature, which become obstacles to an enemy's approach only when fully manned by well-armed, courageous and vigilant troops.

A field work performs all that is required of it, when it compels a force, superior in number to its garrison, to resort to the tedious and costly operations of a siege to take it; or makes this force move in another direction to gain its ends.

The military man should remember that a well-arranged intrenchment, or field work, should obstruct the enemy's approach, as well as give shelter to its defenders. It is a question, oftentimes, difficult to answer, as to which of the two, the shelter to the defence, or the obstruction of the enemy, is the more important. Either one gives an increase of resistance to the defence, and the two combined form an obstacle difficult to surmount.

An enemy may succeed by brute force and sheer numbers, in carrying a position in which the defenders are sheltered simply, or a position arranged with obstacles that hold him under fire at close range; but the success is accompanied by such heavy losses, that the victory, in many cases, had better be classed as a defeat. This is particularly so, where the two are combined.

It is well to remember, that a vigorous defence requires that every part of a field work, or line, should be guarded by a sufficient number of troops to repel any assault that an enemy might make, and that numbers alone are not sufficient. Endurance, courage, and vigilance are necessary in the commander and his troops. No better motto can be devised for the banner of the defence, than the trite aphorism so frequently quoted,

"Eternal vigilance is the price of liberty."

The memoranda, compiled for the use of the officers, commanding the different field works, surrounding the City of Washington, during the war of 1861–5, may be quoted in this connection. The precautions to be taken, and the course advised for each commanding officer, are given in detail and will be as applicable in the future as they have been in the past.

192. Memoranda compiled for the guidance and information of officers serving in the defences of Washington.

"1. The number of men required for the garrison of each work (artillery and infantry supports included) has been calculated, and should be known to every commanding officer of a fort; but it will be for the brigade and division commanders, or for the commanding general, to determine the necessity of filling up each garrison to its full siege complement and of manning the connecting lines of rifle-trenches.

"2. It is the duty of a commanding officer of a fort to see that he has all the means and appliances that may be wanted during a siege provided beforehand. If his position be somewhat isolated and where the enemy may cut off his communications, he should see that he has an ample supply of provisions stored, either in his fort, or in a secure place in its rear; and he must take measures to keep an abundant supply of water in the fort, both for the use of the garrison and to extinguish fires.

"He should see that his fort is provided with all the tools that may be wanted during the siege, particularly with shovels, picks, axes, saws, augers, and hammers. Further, he should provide, either beforehand or as soon as possible after the siege commences, all the materials that may be wanted during the siege, such as an abundant supply of timber, of plank, of nails, spikes, of sand bags, gabions, and fascines. Timber is of the first importance; a large supply of it should therefore be secured.

"3. When any part of the line of defence is threatened with an attack, all houses, trees, bushes, and in general everything that could be used as a cover by the enemy's sharpshooters, should be at once removed for a distance of at least six hundred yards from the line. If sharpshooters can be concealed within that distance of a fort or battery, they will pick off the cannoniers and the guns cannot be served without serious loss.

"In conducting the defence of a fort, each company or detachment, should have its particular post and duty assigned to it, and receive special instructions as to their duties in all possible contingencies.

"Commanding officers should be careful to see that all their subordinates thoroughly understand the ground, both in his front and rear, and are familiar with all the roads and paths; they should know by a careful examination of the ground in front the points which an enemy would select for batteries. and the exact ranges to such points, and should study out how best to arrange their own guns to contend with such batteries should they be erected; they should know the relations of the fort they occupy to the adjacent works, and be familiar with all the resources that may be made available during the defence. This information should be imparted by them to the non-commissioned officers, and in some cases to the privates; for it should be remembered that the fate of the fort may depend upon the good or bad conduct of one individual.

"There should be a reserve in all cases. It should be posted in the bomb-proofs, or behind traverses and magazines, or under temporary shelter made by leaning timber against parapets, magazines and bomb-proofs, or by digging trenches in the ground and covering them with timber and earth.

"Strict vigilance should be exerted to guard against surprise; for this purpose, when the enemy approaches, a chain of sentinels should be posted in front of the works, and as far to the front as practicable, taking particular care to have them posted at all points where an enemy might approach the works without being seen, either for the purpose of reconnoitering or delivering an assault; but the general arrangement of sentinels must be made by brigade or division commanders.

"By keeping up, both by day and night, such a chain of sentinels in front of the works, and by posting, where practicable, a few sharpshooters in holes or rifle-pits in front of each fort, the efforts of the enemy to obtain the information he wants before commencing the siege batteries will be greatly retarded, and time afforded to the defender for completing the preparations to receive him.

"If the sentinels are driven in and an attack is apprehended at night, fire and light balls should occasionally be fired to the front, taking care, however, not to set fire to the abatis; and in this case the guns should be double-shotted, and pointed if possible, by daylight, so as best to sweep the ground within a distance of six hundred yards, depending on the configuration of the surface in each particular case.

"If the ditches are not flanked shells loaded with service charges and ten-second fuzes, should be placed on the banquettes in charge of men specially instructed how to use them. These men should be provided with pieces of burning slow-match.

"Hand grenades, loaded and capped, should also be placed ready for use, and, finally, the men should sleep at their guns and every man should know his post and his duty. Shells and also hand grenades may be exploded by attaching a string in such a way that the act of rolling them or throwing them will, as soon as the *string* is brought *taut*, explode a percussion cap or friction tube.

"4. If the enemy should open his attack by a warm cannonade, and concentrate his fire upon a particular fort, the troops should not be unnecessarily exposed to it, if they can be sheltered near the posts they are to occupy when an assault is made. If the cannonade should become too warm for the garrison to reply without too much loss, the field and siege guns should be removed from their embrasures and placed behind the parapets or, in case of an enfilading fire, behind traverses, bomb-proofs, etc.

"During a cannonade the dead and seriously wounded should be kept out of sight.

"When the assault is made, this cannonade of the enemy must cease in order not to injure his own troops. The guns are then run into position and every man resumes his post. The men should be instructed to reserve their fire until the enemy has arrived at certain points, marked out in front of the works, where it will be most effective.

"Particular attention should be given to securing the gate-way. The abatis in front should be made continuous. As the gates are made to open inward, the enemy cannot easily open them outward. The danger to fear is that he will force them inward, or lift them off their hinges. To prevent the first, roll a heavy log against the inside of the gate, plant posts or drive strong stakes behind it, and, for greater security, cover the log with earth. The second danger may be prevented by "upsetting" the head of the pintle upon which the gate turns, or still easier by driving a strong spike immediately over it. If the gate-way is not flanked it may be necessary, in some cases, to increase the dirt over the log until it becomes an infantry parapet. Communication with the interior may be kept up by one or more light movable bridges, made of plank or boards, leading from the parapet across the ditch. These bridges should be kept inside the fort at night, or when an attack is anticipated.

"5. Every precaution should be taken to secure the magazines against vertical or curved fire. The entrance is the weak point; hence, where necessary, cover it by additional earth; put splinter-proof guards, made of timber or plank, around the door, and have water at hand, in barrels or casks, to extinguish any fire near the doorway; keep both doors shut, especially the inner one, allowing only one or two men to be inside, and only the ordnance sergeant, with an assistant, on the stairway to pass out the ammunition that may be called for. Keep all the ventilators closed, and fill as far as practicable, all ventilating tubes with earth both outside and inside of the magazines.

"See that an ample supply of wads is kept on hand outside of the magazines, and that the ordnance sergeant makes careful report of the ammunition expended, and that it is promptly replaced.

"Fill at once any holes made in the magazine cover. Logs, fascines, or even sticks, laid against its exposed side, greatly reduce the penetration of shot, particularly of elongated projectiles, by deflecting them.

"6. Build merlons between barbette guns, and partially fill wide embrasures, as soon as the positions of the enemy's batteries and the proper direction of fire of each gun are ascertained. Cut away the foot of scarps to render escalade more difficult, taking care not to endanger the stability of the parapet. Use earth so obtained for making a glacis and traverses across the abatis, if it be threatened by an enfilading fire. A few piles of earth across the abatis, particularly if the earth be wet, is a great security against such fire.

"Commanding officers of the forts cannot be too strongly impressed with the fact that the abatis is one of the main sources of strength to a field work. It should be carefully protected from injury and depredation, fire, etc.

"7. Bury percussion shells or hand grenades, to act as torpedoes, in the bottom of the ditch and outside of the abatis.

"8. Put up traverses on all faces liable to be enfiladed, to protect the guns, even if to obtain room for them some of the guns have to be removed; repair all damages to the parapet on the following night, if not practicable to do it before.

"Earth may be obtained for the above uses by excavating in the terreplein for bomb-proof shelters, and by digging pits or holes, about three feet wide and deep, where ricochet shells are most likely to fall, and where the excavations will not seriously interfere with the defence; remove all sheds and wooden buildings lest they take fire.

"9. Construct temporary banquettes on all bomb-proofs and magazines, to afford an infantry fire on the probable front or points of attack. These may be made with plank resting on trestles or posts, or by cutting away the earth so as to afford standing room for infantry, with a parapet in front.

"10. When, or before, the enemy's approaches have been advanced to the vicinity of the work, a surprise, or sudden assault upon one or more of the forts, especially by night, may be anticipated. Under such circumstances it is the duty of the commanding officer, though he may have resorted to every expedient to retard the siege, to impress upon the garrison that they cannot, without loss of honor, either abandon the fort or surrender it without resisting at least one assault. Such is the inexorable law of war, and it should not be forgotten that it holds good whether or not there are other works in the rear to which the garrison might retire with comparative safety. When the attack is about to be made, the commanding officer of a fort should endeavor to inspire his men with confidence in their powers of resistance, with self-reliance and enthusiasm.

"Every preparation should be made to resist the passage of the ditch, by the fire of artillery and infantry, by torpedoes, by loaded shell and hand-grenades, and by attacking the enemy on the parapet.

"The commanding officer must decide beforehand, according to the particular circumstances under which he is called upon to act, how this attack should be made. If there be, behind bomb-proofs and magazine, banquettes and breast-heights for infantry bearing on the points of attack, then, at the proper moment, the field-pieces and howitzers should be withdrawn from the platforms, loaded with double charges of canister and placed in positions where they may still be used against the enemy as he appears on the crest of the parapet or descends to the terreplein; while the infantry should line these interior parapets and aid, by their fire, in driving the enemy from the work, the reserve charging furiously with the bayonet on the appearance of any confusion or disorder.

"If there be no such interior lines of infantry fire on the point of attack, then the enemy must be met with the bayonet on the top of the parapet. For this purpose one or more steps, made of boards, should be prepared beforehand, to enable the infantry to mount the parapet, and it should be impressed on the defenders that the assaulting troops, arrived at position, will be greatly fatigued, and necessarily in disorder, and, moreover, will be cut off from all external support, and hence that it is an opportune moment by a vigorous assault with the bayonet to hurl the enemy into the ditch and to retrieve, with disaster to the foe, the endangered possession of the work and of the defensive line on which the safety of the nation depends.

"The firing of the guns, particularly howitzer and field-pieces, loaded with canister, should be continued as long as possible in order to delay the advance of the supporting columns; and, if the assault fail and the support be driven back in disorder, a sortie in force may be made and the enemy pursued into his works.

"Such sorties, however, should not be attempted by the troops in any particular fort, but should be made, under the direction of the commanding general, by the outside reserves, supported by the fire of all the forts and batteries bearing on the position to be taken."

193. Examples.—The most noted examples in recent wars, of intrenched positions, in which the works used were field fortifications, and which illustrate the attack and defence, are found in the cases of Sebastopol in 1854, Vicksburg and Port Hudson in 1863, Richmond and Petersburg in 1864, Plevna, in 1877–8, etc. The student is referred to these particular examples for the details which cannot be given here.

From the many examples, reference will be made here only to the assault made upon Fort Sanders in 1863.

After the battle of Chickamauga, General Longstreet was sent into East Tennessee to capture Knoxville, then occupied by the United States forces commanded by General Burnside. The latter succeeded in intrenching his position, and held the Confederate troops at bay.

After the victory at Chattanooga, General Grant sent troops to relieve General Burnside, and General Longstreet knowing of their approach determined to risk an assault upon Burnside's position.

He selected the key point of the position, as the point of attack, which was defended by an unfinished earth work, laid out under the direction of Captain Poe of the United States Engineers.

There had not been time to make an abatis, to build powder magazines, or even to revet the work as it should have been. A wire entanglement had been arranged in front of the work, fastened to the stumps of the trees that had been cut down to clear away the ground.

At dark, on the 28th of Nov. 1863, the Confederate sharp-shooters were pushed forward to within rifle range of the line. At half past six in the morning of the 29th, a heavy fire of artillery was opened on the fort and lasted for twenty minutes, or half an hour.

This fire ceased, being replaced by a fire of musketry, and a rush of a brigade of infantry upon the salient of the work.

The wire entanglement tripped many of the assailants, who fell upon the ground, but numbers of them succeeded in entering the ditch and attempted to assault the parapet. The heavy fire of the defence from the flanks and from the parapet, drove them back and, the assault failed. General Longstreet then withdrew his forces.

The readiness, with which the assault was met, prevented its success.

194. The history of attacks made upon field works in recent wars, shows that assaults upon them were generally unsuccessful, where a vigorous defence was made; and when successful they were accompanied with great loss of life, and injury to the assailant.

The conclusion may be fairly drawn that a field work, vigorously defended, cannot be captured by *assault* if it has a good profile and is provided with the accessory means of defence; or if it is commanded by other works which can not be assaulted; or, if the garrison can be readily reinforced from a strong body of troops within supporting distance.

A vigorous defence supposes vigilance and security against surprise. An attack may be, nevertheless, an unexpected one, and may task all the efforts of the defence to repel it. The success of an assault is greatly dependent upon the suddenness with which it is made, and a commander should remember,

> " When 'tis done, then 'twere well
> It were done quickly."

CHAPTER XVI.

SIEGE WORKS.

195. Siege Operations.—Fortified positions, vigorously defended, can rarely be taken by assault. As a rule, positions of this kind can be carried only by the attacking force getting near enough to the defenders to overpower them by the actual contact of superior numbers.

The assailant, to profit by his numerical superiority, must remove or overcome the obstacles between him and the defence. To be able to remove these obstacles, he must protect himself as much as possible from the fire of the defence, so that he may get near enough to the obstacles to overcome them. This moving forward "under cover," and removing the obstacles in his way, make the advance of the assailant a slow one. The attack necessarily becomes a protracted one, and receives the name of **siege**. (Art. 182.)

The labors and movements of the assailant, by means of which he gets near enough to the defence, occupying a position of this kind, to render an assault practicable, are known as **siege operations.**

196. Investment.—The final success of a siege is greatly dependent upon the isolation of the defence from all aid. This isolation is effected by cutting the communications of the defence, and making it impossible for him to receive aid, either by reinforcements, or by supplies.

This interruption of the communications and the isolation of the defence form the **investment of the position.**

The investment is usually performed by a strong body of troops detached from the attacking force, which body moves quickly and suddenly, surrounding the position, and seizing all the avenues of approach.

A chain of outposts and sentinels, placed just outside of the range of fire of the defence, but close enough to watch all the avenues leading to the position, is established by the investing force. This chain is drawn in nearer to the position at night, and moved back a short distance in daytime. The terms **nightly cordon** and **daily cordon** are frequently used to designate this chain.

197. Posting the attacking force.—The main body of the attacking force follows closely the investing detachment, and takes a position in supporting distance of the cordon.

The question is then decided as to the character of the attack which is to be made, that is, whether an assault is to be tried, or whether a protracted attack is to be made. In the latter case, whether the attack shall be a simple **blockade,** a **bombardment,** or a **regular siege.**

A **blockade** consists in surrounding a position and preventing supplies from entering the place, until the defenders, driven by want, are obliged to surrender.

This is a slow process, but effectual in many cases, especially when the nature of the surrounding country enables the assailant to make the investment complete, and when the troops defending the position have scanty supplies.

A **bombardment** consists in directing a heavy fire of shot and shell upon the position, destroying the defences, the magazines, supplies, etc., and wearing out the powers of endurance of the defence.

A **regular siege** consists in approaching the position under cover, removing all obstacles in the way of an attack, and making an assault upon the defence.

A complete or systematic siege of this kind is undertaken when the position cannot be carried by assault, and when the methods of blockade or bombardment are too slow or too uncertain. These latter methods are both used, however, in conjunction with the regular siege, if possible; and, in the cases used to illustrate siege operations, it is supposed that the investment is complete, and that the necessary arrangements for a bombardment are made.

A regular siege being determined upon, the general in command of the attacking forces designates the points exterior to the position which are to be occupied by the besieging forces. The camps are then laid out, being placed beyond the range of the heaviest guns of the defence, and, as far as practicable, upon sites favorable to the health and comfort of the men.

Good communications are at once established between all the camps, and all obstacles, which would impede the free circulation of the troops from one part to another of the ground occupied by them, are removed, or modified. Every precaution should be taken to prepare the ground so that a rapid concentration of the troops can be made, whenever it may be necessary. Strong field-works should be constructed in the vicinity of those camps exposed to attack.

198. Lines of **countervallation**, and of **circumvallation**.—The line of field-works constructed in front of the camps, and on the side next to the besieged position, to defend the camps, parks, and trains against attacks which might be made by the besieged, is called a line of **countervallation**. Formerly, a line of works placed on the opposite side of the camps was used to prevent small detachments from slipping through the lines of the besiegers, and going to the aid of the besieged. This line was called a line of **circumvallation**. Such a line is not often used at the present day, because of the greater mobility of modern armies. A line of this kind might be employed as a line of defence against a relieving army. Even in this case—the advance of a relieving army—it would be better not to wait passively within the lines until attacked, probably by both the reliev-ing forces and the besieged at the same time, but to move out and meet the relieving forces at some other point.

When these lines are used, circumstances will decide as to the kind, whether they shall be continuous lines, or lines with intervals, etc. The activity of the defence, the nature of the ground, the strength of the relieving army, etc., are all factors of the problem which is to be considered.

The question of moving out to meet a relieving army involves, frequently, the question of raising the siege. If the besieging army is strong enough to permit it, a force is usually detached to watch the movements of the relieving army, while the main body remains prosecuting the siege operations. This detached body is known as an **army of observation**.

199. Preliminaries.—There are many things to be attended to and procured before the actual labors of the siege can be vigorously and systematically prose-cuted. If the siege is to be a vigorous one, these things must not be neglected. Principal among these, are the location of the parks and trains in convenient and secure places; the construction of magazines for am-munition, supplies, etc.; the supply of necessary tools and implements; the supply of materials to be used in the siege works; the preparation of these materials; the complete reconnoissance of the fortifications defend ing the position and the ground immediately in their front, etc.

All these preliminaries should be attended to and provided for, if practicable, while the main body is arranging the camps and intrenching the ground to be occupied.

200. Selection of the side of attack.—The general, in selecting the side upon which to attack, tries to choose that portion which, being gained by him, forces the defenders to surrender, or to retreat, if there be any way for them to retire. That point whose occupation compels an abandonment of the position, or its surrender by the defence, is known as the **key-point.** Its selection proves the skill and fitness of the general for the command entrusted to him.

Previous to an advance upon the position, the general has acquired more or less information about the defences, the strength of the defenders, the nature of the ground, etc. Upon reaching the spot, reconnoissances are made to verify, or correct, this information, and add to it in every way.

These reconnoissances are supplemented by instrumental surveys, and by other means, whose object is to determine the exact distances, and the true directions of the lines of the defence, the general features of the ground, and the kind of obstacles to be encountered.

The general has to decide from this information thus gained, viz :

1. Which part of the position is easiest to carry;

2. Which part carried gives possession of the rest; or, which part is the *key-point;*

3. Which side of the part selected is the best on which to make his approaches; and,

4. Which part selected would be the best, taking into consideration the establishment of his depots and lines of supply, and the probabilities of an attempt to relieve the besieged.

These questions are partially answered before the posting of the besieging army is completed, as it would be bad policy to have the troops encamped too far from the ground where the main operations of the siege are to be conducted.

201. First parallel.—It does not fall within the scope of a work like this treatise to give the details connected with siege operations. These details are fully given in the Manuals prepared for Engineer troops, and for the Artillery. But as every officer of the army in actual service is liable, at times, to be employed more or less upon works of a similar nature, it is thought advisable to describe the simpler portions which may fall to his lot to execute.

The portions of the defences which are to be carried having been selected, the general gives instructions to make preparations for beginning work upon the batteries and first parallel.

The **first parallel** is a **simple trench**, such as

described in Art. 171, and is intended as a protection for the infantry line which is drawn up nearly parallel to the side or **front** to be attacked, hence its name. It also affords a secure base for nearer approaches upon the position.

Before the improvements in modern artillery, the construction of the first parallel and of the covered communications leading from it to the depots in the rear, formed the first steps, when breaking ground for a siege. As soon as it was dark, the working parties were posted just out of range of grape-shot, and were protected by strong detachments of troops under arms. The work was immediately begun, and was practically finished by morning. Even if discovered by the besieged, no particular interruption was expected, because of the inaccuracy of the fire at night, and, the distance of the working parties from the works. Under the protection of the troops placed in this parallel, the batteries, if any were to be used at this period, were begun on the second night, and the approaches were pushed forward.

It is usual now to establish batteries before, or simultaneously with, the construction of the first parallel, so as to engage the artillery of the defence, and keep down its fire. Otherwise, the working parties would suffer great loss from the artillery fire of the besieged, and would not be able to finish their tasks by morning.

Batteries armed with heavy guns are, in some cases, constructed at distances of from one to three miles from the works to be attacked, and, at a specified time, these guns open their fire upon the besieged. Field batteries are also, under cover of darkness, pushed close in, and add their fire to the others. Skirmishers move in close to the position, covering themselves in rifle-pits and shelter-trenches, and keep up a warm fire on the besieged. All this is done before the commencement of the first parallel, and for the purpose of testing the endurance of the besieged. In the meanwhile, preparations are made to begin work upon the first parallel, and if the bombardment fails to force a surrender, the work is begun upon the parallel.

202. Construction of first parallel.—The position of the parallel should be marked or traced upon the ground before the time of beginning the work, care being taken to have no marks which would attract the attention of the besieged. This tracing upon the ground is executed by engineer soldiers, and in accordance with the methods given in the manuals of engineering.

The working parties, detailed from the infantry, are brought to a convenient spot near the place, and furnished with picks and shovels. It is recommended that they be divided into reliefs.

On the arrival of the engineer officer reporting the tracing completed, the working parties move forward to the ground to be occupied, being led by this officer. The first relief is marched off in fours, by file, or in column, according to the circumstances of the case. Each man carries a pick and a shovel, and slings his piece so as not to interfere with his movements. Upon reaching the ground, the men are extended along the line, indicated by the tapes, which marks the direction of the parallel. Engineer soldiers assist them in this extension, and show the men their places and their tasks. The men are placed five feet apart. Each man drives his pick into the ground, on the left of his task, lays his shovel in front of him, behind the tape, unslings his musket, takes off his waist-belt, and gets ready for work. At the proper command, he begins to dig.

The men are instructed to keep silence, and to allow no clashing of tools, or of arms. All words of command are to be given in a low voice ; lights and smoking are not allowed.

Lanterns of a peculiar shape are allowed to the engineer soldiers, who use them in tracing the line marking the parallel.

The men begin their work as soon as they are told. Each man digs into the ground, making a hole about three feet in circumference and four feet deep, throwing the earth in front of the tape, and at a distance from him equal to the length of his shovel, which is a little over three feet. He then widens the hole to the limits assigned him.

Fig. 91 shows the profile of the parallel excavated by this method, known as the simple trench.

The space, A, shows the portion which is to be excavated by the first relief, who, having executed the task assigned it, returns to camp, leaving the tools upon the

Fig. 91.

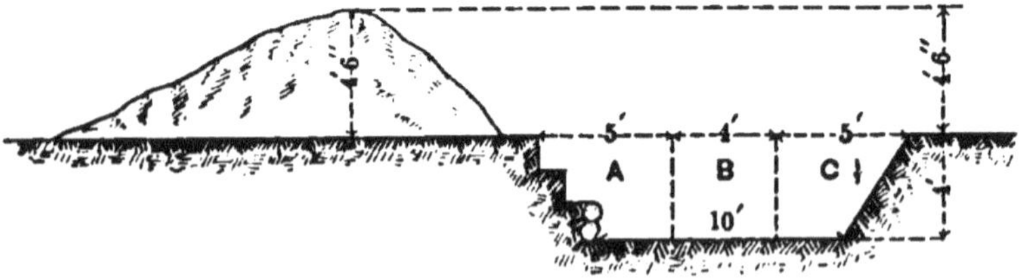

ground. The second relief executes the part marked B. The third relief completes the part marked C, and constructs the lower step, using fascines, as shown in the figure.

In the days of smooth-bore pieces, this first parallel could be executed within 600 yards of the salients of a fortified place, this being outside of the range of grapeshot; but in these times, and in the presence of an active enemy, well supplied with arms, the distance must be much greater. This distance will depend upon the features of the ground, which always afford more or less cover to the working parties, as well as upon the nature of the defence.

The ground to be occupied by the first parallel may be obtained by pushing forward skirmishers, who intrench themselves as they advance. Having gotten possession of the ground, instead of building the parallel in a single night, the parallel may be constructed by throwing up trenches to connect the rifle-pits and shelter trenches occupied by the skirmishers. In this case, a strong body of covering troops should be used to guard the workmen from the sorties of the besieged.

The length of the parallel must be sufficient to embrace the front of attack, and the parallel should have its extremities protected by strong field works.

203. Approaches.—The **approaches**, or **boyaux**, as they are sometimes called, are trenches whose general direction is towards the works to be captured. They have a **zigzag** direction to prevent being swept by an enfilade fire, the prolongation of each branch passing outside of any work of the defence which is in range.

The profile of an approach is shown in Fig. 92.

Fig. 92.

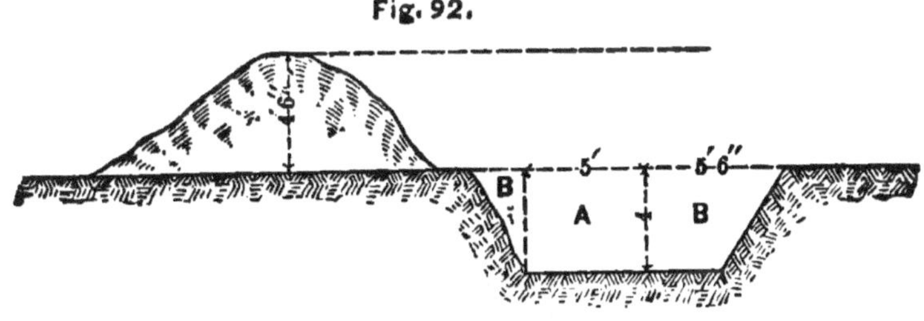

Each branch of an approach overlaps that behind it by about ten yards, to afford protection against enfilade, and to serve as a trench-depot for trench-materials.

The dimensions may be the same as those used in the first parallel, and frequently these are adopted, if there is to be much use made of the approach. In the figure, the approach is made only nine feet wide at bottom. Its construction, when it is a simple trench, conforms to that already described for the parallel, and can be understood from the figure.

In both the parallel and the approach, it is recommended to slope the bottom of the trench to the rear, giving a fall of about six inches. This provides for drainage, and also affords greater protection to the men using them.

It will be observed that steps are not indicated in the approach, like those shown in the parallel. They are used in the latter to allow the guards in the trenches to fire over the crest of the parapet. If it be required to post troops in the approach, or have men fire from it, steps should also be placed in it.

204. Flying trenchwork.—It is not always practicable to use the simple trench in these constructions, in consequence of the exposure of the working parties to the enemy's fire. A more expeditious method of obtaining shelter is adopted. The method used is known as the **flying sap**, or better, **flying trenchwork.**

Fig. 93.

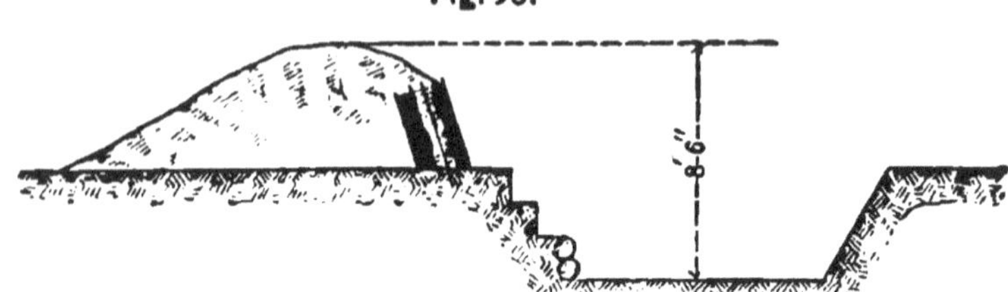

This method consists in placing a row of gabions along the front of the tracing tape, and filling them as quickly as possible with the earth excavated from the trench. When the gabions are filled, the rest of the trench is excavated, and the earth thrown over and behind the gabions, thus forming a parapet. (Fig. 93.)

By this method it is seen that shelter is obtained much more quickly than by the method of the simple trench.

205. Second and third parallels.—Tactical considerations require the construction of other parallels, as the besiegers approach nearer the work. There should be at least two more, making three in all, and it may be necessary to use a greater number. The French, at Sebastopol, found it necessary to use seven.

The second should be near enough to the first to have the workmen constructing it under the protection of the troops posted in the first parallel. This second parallel is within close artillery range of the defences, and nearly within the zone of accurate fire of small arms. It must, therefore, be executed by the method of "flying trenchwork."

From this parallel the work has to be carried on by trained engineer soldiers.

The *third*, or last, parallel is so close to the work that there is a prospect of the besiegers reaching the parapets of the defences by making a rush from the parallel. This distance to be rushed over is usually assumed to be not more than about thirty yards.

Assuming the position to be fortified by a bastioned

work, F, (Fig. 94) with out-works, A, B, C, etc., the relative positions of the parallels and the approaches can be seen in the figure.

Fig. 94

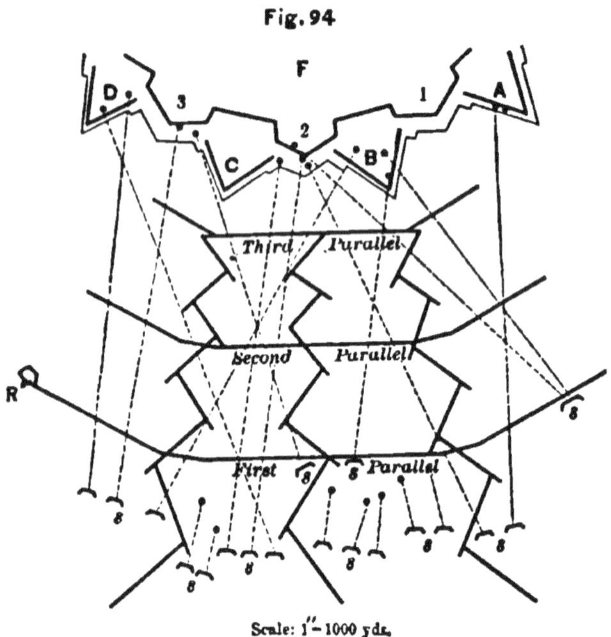

206. Batteries.—The term **battery** is usually applied to a collection of guns; it is also used to designate a prepared position, in which the guns and gunners are covered wholly, or in part, from the fire of the enemy.

Batteries are divided into classes, according to the kind of fire used, as **embrasure**, or **barbette**; according to the object to be attained, as **enfilading, counter**, and **breaching** batteries; according to the kind of gun used, as **mortar, howitzer, rifled** batteries, etc.; according to the position of the terreplein, as **elevated**, **sunken** batteries, etc.

The batteries constructed and armed in the beginning of the siege, which open their fire before, or simultaneously with, the beginning of the first parallel, are placed within effective range, and so as to bring direct and enfilading fires upon the defences. The location of these batteries is a matter of judgment and experience, care being taken to have them effective, and yet not interfering with the execution of the first parallel.

For convenience of supply, the batteries may be located in groups, but not so close to each other as to form a target upon which the enemy can concentrate his fire. An effort should be made to construct them unnoticed by the besieged, and hidden as much as possible from his view, so that, even after they have opened their fire, it will be difficult for him to distinguish their positions.

It is usual to have not more than four or six guns in any one battery. A greater number forms a larger target for the enemy's fire, and is apt to cause the besieged to concentrate a stronger fire upon the battery.

The batteries are distributed over a wide space to prevent concentration of the enemy's fire, and at the same time to allow the besiegers to bring a strong converging fire upon the defences.

They are placed so as to bring strong direct and enfilading fires upon those faces and parts of the defence which sweep with their fire the ground to be occupied by the approaches.

The distance of the *first* artillery position from the defences, and the inaccuracy of the fire of these batteries, will make it difficult to silence the guns of the besieged which bear upon the ground over which the attack is to be made. A position nearer to the defences must therefore be occupied. This *second* artillery position is taken as soon as the first parallel is completed.

The batteries in this second position will be armed partly with new pieces brought up from the siege-parks, and partly with the guns taken from the batteries of the first position. With the increased range of modern artillery, these batteries are effective when placed in rear of the first parallel, and will, in future, generally occupy positions like those shown in Fig. 94, as indicated by the letters S, S, S, etc. The dotted lines shown in the figure from a few of the batteries, show whether they are enfilading or counter-batteries, and it is seen that these batteries are situated so as not to have their fire interfere with the approaches. Other batteries, not shown in the figure, bring their fire upon the front of attack.

In addition to these batteries, mortar-batteries, and batteries for guns firing at high angles, are used to throw shells into the interior of the defences. And if masonry walls or revetments are to be destroyed, breaching batteries will have to be used. When the masonry or revet-

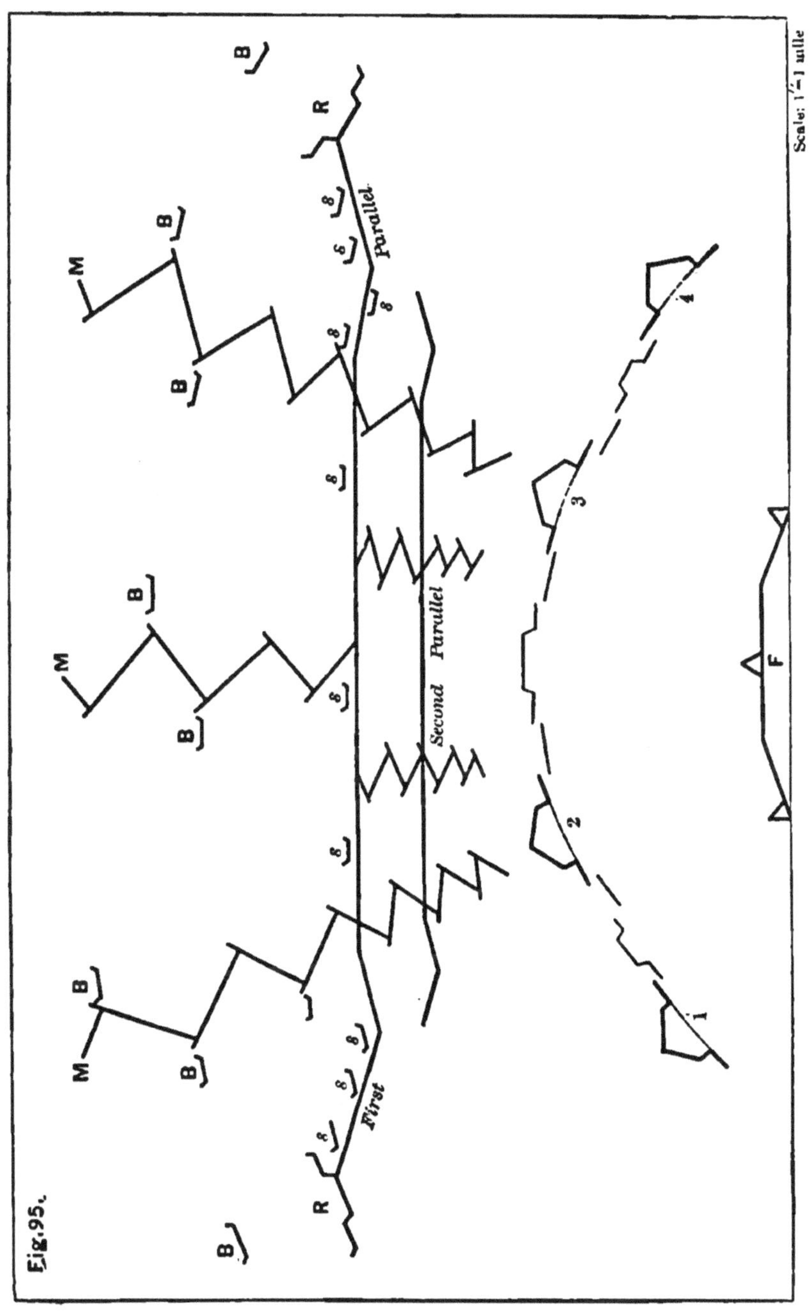
Fig. 95.
First
Parallel
Second
Parallel
R
M
B
F
1
2
3
4

ments are protected from distant fire, the guns of the breaching batteries are used as enfilading or counter-batteries until emplacements can be obtained within breaching distance.

As further illustrative of the relative position of the batteries and parallels, suppose an attack to be made upon a position defended by a line of detached works, with an interior work F, as shown in Fig. 95, these detached works being connected by covered communications (Art. 171, Fig. 83) and by shelter-trenches. The letters M, M, etc., indicate the covered communications connecting with the magazines and depots of trench-materials; B, B, B, etc., indicate the batteries of the *first* artillery position; S, S, S, etc., the batteries of the *second* artillery position; 1, 2, 3, etc., the detached field-works forming the outer line of defences, etc.

207. Construction of batteries.—The details of construction are given in the manuals.

Some few general principles may be enunciated as governing their construction.

Batteries in which the platforms of the guns are on the natural surface of the ground, or above it, are known as **elevated batteries**. Those in which the platforms are below the natural surface are designated as **sunken batteries**. The latter are of course more quickly built.

Batteries in which the gun-platforms are laid on the natural surface of the ground are more simple in construction than any of the other kinds, because the arm-ing of the batteries can be made independent of the execution of the parapet.

Enfilading batteries should have their crests as nearly as possible perpendicular to the prolongation of the line to be swept by their fire; counter-batteries should have their crests nearly parallel to the line subjected to their fire. These conditions, when fulfilled, will save a great deal of labor in the construction of embrasures.

Embrasure batteries are to be preferred to barbette, but the embrasures must be hidden, if possible. Various devices are given by means of which the embrasures are concealed, and the gunners screened from the fire directed upon them.

The general principles governing the construction of field-works and trenches apply to the construction of batteries. It is to be remembered that artillery fire draws artillery fire; hence a greater thickness is requisite for parapets and epaulements sheltering artillery, than is required for the simpler works.

The batteries forming the first artillery position may have weaker profiles than those of the second position.

208. Remark.—The works in advance of the second parallel are executed by the **saps**, the workmen being engineer soldiers, known as **sappers**. In operations where it is only intended to get near enough to make an assault practicable, it may be possible to carry on the work by means of the **single sap**. It is probable.

however, that either the **full**, or the **double sap**, may have to be used, and possibly both of them.

The last parallel—the third, if there are only three—as has been stated, is so near the defences that the probabilities are on the side of a rush being successful, so far as the assailant being able to reach the ditches in front of the works, is concerned. If it is decided to make the assault from the third parallel, this parallel is arranged with steps to allow the troops to rush over the parapet, at the command. The assault is then made, like that already described for taking a field-work. A heavy artillery fire is opened to drive the defenders from their parapets and, under its cover, the assaulting columns, preceded by pioneers and engineer soldiers to remove the obstacles, make a rush for the defences, dash over the parapets, and overpower the defenders by a hand-to-hand conflict.

If the ditches are revetted with masonry, or if there are obstacles which render an assault impracticable, siege operations must be continued.

Pushing forward the approaches, establishing breaching batteries, battering down the walls, crossing the ditches, etc., are part of the operations. This part of the siege is more laborious, and far more dangerous, than that which preceded. All the works are planned and executed under the immediate direction of engineer officers, and by trained engineer soldiers. The methods pursued are given in treatises upon the "attack and defence of permanent fortifications," and the details are laid down in the manuals for engineer troops.

The defence of a position which is besieged is practically the same as that for a field work. The memoranda given on page 228 are applicable to the case.

209. Maxims of Vauban.—Marshal Vauban, the great military engineer, laid down certain general principles which he recommends to the profession for its guidance in conducting siege operations.

It may be as well to state what is told of Marshal Vauban, that he was present at, and conducted, fifty-eight sieges. An experience of such magnitude as this enabled him to formulate these maxims, which are accepted, even now, as good authority, for matters of this kind.

VAUBAN'S RULES.

1. To delay opening the trenches until the besieging forces are all well posted, and until everything requisite for carrying on the siege vigorously has been collected.

2. A single, rather than a double, attack should be preferred, unless the two can be well connected, and unless the besieging force exceeds the garrison considerably in strength.

3. The parallels and approaches should embrace all the defences which bear upon the site to be occupied by the besieger's works, so as to secure positions for estab-

lishing the batteries that may be required to silence the fire of these defences.

4. The greater the number of approaches, the better will they be for giving mutual support, less encumbered communications, and for dividing the fire of the defences, which, if concentrated upon a single one, might soon destroy it.

5. There should be at least three parallels, placed in the best positions for mutual support, and to guard the approaches and batteries from sorties of the besieged.

6. Never, if it can be avoided, attack a point upon which the approaches can be run only on a narrow front; nor one which can only be approached over marshy ground, or on causeways.

7. Be careful not to push forward any portion of the trenches, until they are well flanked and protected by trenches in their rear, which are completed and occupied by troops.

8. Avoid encumbering the approaches with trench materials, tools, workmen, or troops; these should be kept in the parallels, on the right and left of the approaches, so as to be at hand when wanted.

9. The ricochet batteries should occupy positions such that they can have enfilading and slant reverse fires upon the guns of the defences.

10. Fire should not be opened from any series of batteries, until it can be done at the same moment from all of them.

11. The fire of the batteries and trenches, rather than open assaults, should be used to drive the besieged from their defences, before attempting to occupy them by the besieging force.

12. When it is decided to make an open assault, it should be made in day-light, if the fire of the defences, which bears upon the point to be carried, is completely kept under by the fire of the batteries and trenches; but, when the fire of the defences is not completely kept under, the assault should be made during the night.

13. In offering resistance to an open assault of the besieged upon any unfinished portion of the trenches, it is suggested to withdraw the workmen and the few troops near them, to some point within the parallel immediately in the rear, and then to repel the assault by opening a vigorous fire upon the assailing force.

14. The defence should keep within the cover of the parallels while the assailant is advancing to the assault, leaving him to expose himself to the fire of the defence until he is cut up, and is in confusion in the trenches that he may have carried; then attack with the bayonet and drive him out.

15. Attacks of this kind should not be pushed too far, but should cease in time to allow the troops to regain shelter within the trenches, before the besieged can open fire from the works.

CHAPTER XVII.

MILITARY BRIDGES.

210. Communications.—In speaking of the **communications** of an army, common roads, railroads, navigable rivers, canals, and telegraph lines are usually meant, as it is by means of these that the different parts of an army, in its military operations upon land, are connected with each other.

The establishment of these communications, and their maintenance, give the general in command of an army great solicitude. So important are they in military operations, that they largely influence the general in his selection of the theatre of operations, if the choice is left to him.

The establishment and maintenance of these communications form a part of the duties ordinarily assigned to the engineers, and the construction of bridges plays a most important part of this particular duty.

211. Passage of rivers.—An army moving forward oftentimes finds its march interrupted by a large stream, or river, intersecting the general line of advance.

The army may be crossed over either by fording, by ferrying, or by bridging the stream. Which of these methods should be adopted will depend upon the depth of the stream, its width, the character of the bottom, the strength of the current, and the means at hand.

212. Fords.—The requisites of a good ford are, small depth of water, low banks, moderate current, and hard bottom.

A ford, practicable for small bodies of troops, oftentimes becomes useless for large ones, because the bottom becomes stirred up by the passing troops, and is carried off by the current, and the ford finally becomes too deep to be used.

Fords may be discovered by examining the stream and its banks. Paths which lead to a stream and which come out on the opposite side indicate a point of crossing; and an examination will show if it is a ford, or if other means are used to cross at that point.

When there is a probability of the water being shallow enough to admit of fording, mounted men may be sent in to test its depth and fitness of the bottom and banks. Men, in boats, can easily ascertain the depth by sounding.

The water should not be deeper than three feet for infantry, and four for cavalry.

Fords, when used by large bodies of troops, should be marked out, and, if deep water is near the place of crossing, boats should be stationed to assist men who may be washed down the stream by the current. The force of the current may be broken by stationing mounted men in the stream, above the line of crossing.

Fords should be examined after freshets, to note any changes which may have occurred.

Fords with rocky bottoms formed of rolling stones, boulders, etc., can with difficulty be used by troops, and are almost impracticable for wheeled vehicles.

Fords may be destroyed by digging trenches across them, or by obstructing them with obstacles of such a nature as to impede the progress of the men, or animals that attempt to use them.

213. Ferries.—Ferrying is usually done by means of boats. But when boats cannot be had, or when there are not enough of them, the ferrying may be done by using rafts, or some other buoyant arrangement.

The boats and rafts may be propelled by oars, they may be drawn across by ropes, or they may be made to move by the action of the current.

The simple **rope ferry** is frequently seen in use over streams of moderate width and with a sluggish current. A rope is stretched from bank to bank, and men, standing on the raft or in the boat, seize the rope by their hands and pull the boat along.

When the boats are made to cross by the action of the current, the method is known either as the **trail** bridge, or as the **flying** bridge.

In the trail bridge, a rope is stretched across the stream, and drawn very tight, to keep it above the water. The boat is attached to this rope by a pulley, and is made, by means of its rudder, to have its side make an angle of about fifty-five degrees (55°) with the direction of the current. The force of the current acting upon the side of the boat may be divided into two components, one parallel to the rope, and the other perpendicular to it. The latter component is balanced by the connection of the boat with the rope; the other component drives the boat across, the pulley allowing motion in that direction. This method requires a velocity in the current of not less than three feet a second, and a width of stream not greater than one hundred and fifty yards.

The flying bridge is employed when the width of the stream is too great for the use of the trail bridge.

The principle of the flying bridge is the same as that of the trail bridge. The difference is in the details employed.

In this bridge, instead of stretching a rope across the stream, a cable is used, one end being fastened to the boat, or raft, and the other end anchored in the stream. The cable is supported at intermediate points by small boats, casks, or other means, to keep the cable above the surface of the water for the necessary distance.

All these methods of crossing a river are frequently used, and possess peculiar merits. But when large bodies of troops with their transportation are to be crossed, they do not offer the advantages of the bridge.

214. Military bridges.—A **military bridge** is a structure erected over a water-course to afford a continuous roadway between the opposite sides of the stream which can be used by troops in crossing from one side to the other.

Military bridges are of two general kinds: 1. Those in which the roadway rests upon *floating* points of support; and 2. Those in which the points of support are *fixed.*

Military bridges are ordinarily but temporary constructions, built to serve a given purpose, for a limited time. Those built as an army advances are usually constructed of materials which can be speedily collected at the point required, from the supplies found in the vicinity, or from those which have been transported with the army in its advance. Those built in rear of an army may be more durable, and may, in some cases, be permanent in their character.

An enemy retiring before an army, or contesting its advance by defensive operations, either destroys the bridges and the materials from which they may be built, or defends them so that they cannot be used by the advancing forces.

To provide for such contingencies, it is the custom, in well equipped armies, to carry with them a "bridge equipage," by means of which a bridge can be constructed across a stream in a short time, so as to allow the army and its trains to pass over without delay.

215. U. S. Bridge Equipage.—A history and description of the bridge equipage of the United States army are given in a book published by the Government, and entitled the "Organization of the Bridge Equipage of the United States Army, etc."

This bridge equipage is so arranged that it may accompany an army, and, when needed, may be used to build a bridge of sufficient strength and stability to pass the army and all its trains with safety over the widest rivers; or it may accompany an advance guard, a cavalry expedition, or other small force, and be used to afford a passage to the body which it accompanies. The former is known as the **reserve equipage**, and the latter as the **advance-guard** equipage.

In both of these cases the bridges are constructed with **floating** supports, although **fixed** supports may be used, in connection with the others.

216. Ponton bridge.—The bridge constructed from the materials of the "bridge equipage" is known by the familiar name of **ponton bridge**, the points of support in the water being called pontons.

The ponton belonging to this equipage is a boat thirty-one feet long, two feet and eight inches deep, five feet and eight inches wide at top, and four feet and five inches wide at bottom. (Fig. 96.) These are the outside measurements.

The bow and stern are built a little higher than the body of the boat; the width of bow being two feet and nine inches, and of the stern, four feet and eight inches.

A boat of these dimensions, built in accordance with the specifications required by the government, as laid down in the book before mentioned, will have capacity sufficient to transport forty men, armed and equipped,

Fig. 96.

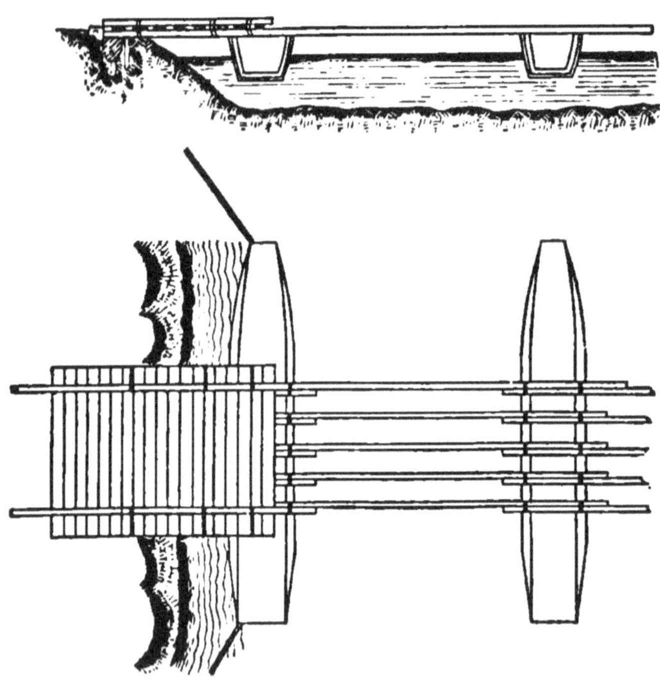

besides its complement of pontoniers, seven in number, and will be strong enough to sustain the loads which pass over the bridge.

The beams, known as **balks**, which are used to support the flooring of the roadway, are of white pine, and are twenty-seven feet long, with a cross-section of five inches by five inches. A small cleat of oak, called a

claw, is fastened on the under side at each end. The distance between the claws is twenty-five feet and eight inches. The boards, known as **chess**, forming the flooring, are white pine planks, thirteen feet long, twelve inches wide, and one and one half inches thick. The width at each end is reduced to ten and one half inches, for a distance of two feet, to allow room for the lashings of the side rails.

Side rails are balks placed on the flooring, and lashed firmly to the balks on which the chess rest. The rope used is one inch manilla rope, and is called **rack lashing**.

The method of combining the boats, the balks, etc., is shown in the figure. A bridge made of these boats, and as indicated in the figure, is strong enough and stiff enough for all the purposes of passing an army and its train.

217. Building the bridge.—There are several methods used for building bridges with this equipage. The one most generally employed is known as the construction by "successive pontons," and is as follows:

The place for building the bridge having been selected, the pontons are brought to the banks of the stream, near the spot, and the boats are launched into the water. Each boat is provided with an anchor. Some of the boats cast their anchors up-stream, while others cast them down-stream. The number of anchors to be cast will depend upon the rapidity and strength of the current. Under ordinary circumstances, an anchor cast up-stream from every alternate boat, and half the number down-stream, will be sufficient.

The boats casting up-stream anchors are launched above the bridge; the others below the bridge.

If none exists, an easy approach for the wagons and artillery should be constructed, leading down the bank to the bridge. A strong sill is then imbedded in a trench, perpendicular to the axis of the bridge, and is held firmly in place by four stout pickets, driven about eight inches from each end. This sill is horizontal, and should be as nearly as possible on a level with the flooring of the bridge.

A ponton is then brought up opposite to this sill, and close to it. Five balks are brought forward, and the ends placed upon, and lashed to, the outer gunwale of the boat, in the proper places. The men holding the balks push the ponton off, until the ends of the balks on shore rest upon the abutment sill. The ponton is then secured in position by shore-lines running out from the bow and stern, and fastened to mooring-pickets. The chess are brought forward and laid upon the balks, to within one foot of the boat. A second ponton is brought alongside of the first; five balks are again used, and this second boat pushed out. The balks are firmly lashed together and to the gunwales of the first ponton. The intervals between the pontons are known as **bays**.

The chess are laid as soon as the balks are lashed, and when a bay is completely covered, the side rails are laid and lashed to the balks beneath.

This opèration is continued until the entire length of bridge is obtained.

It is recommended to strengthen the first bay by using two additional balks—one between the first and second, and in contact with the latter; the other, between the fourth and fifth, in contact with the fourth.

When the water is not deep enough to float the first ponton, a trestle, or other fixed point of support, may be used instead of the ponton.

218. Canvas ponton.—The great objection to the ponton just described is its weight, which makes its transportation over bad roads difficult. For bad roads and rapid movements a lighter ponton than this wooden boat has to be used. The one employed under these circumstances is the **canvas ponton**, which consists of a wooden frame covered with canvas.

The wooden frame comes apart, so as to be easily loaded on wagons for transportation. It has two side frames, trapezoidal in shape, the upper piece being twenty one feet long; the lower, eighteen feet and four inches long. The depth of this frame is two feet and four inches.

The frames are connected by pieces called **transoms**, framed into the side frames, and these latter are fastened together by ropes passing through rings in the ends of the frames. The inner width of the boat frame, or distance between the side frames, when the parts are arranged, is four feet and eight inches. In some cases, the side frames are hinged in the middle, so that when taken apart, they may be folded up.

The canvas cover is made of cotton duck. The balks are twenty two feet long, with a cross-section of four and one half inches, and are provided with claws as before described. The chess is the same as that described, but only eleven feet long, instead of thirteen.

219. Organization.—The reserve equipage is divided into trains, each train being composed of four ponton divisions, and one supply division. Each ponton division contains all the material necessary to construct a bridge of eleven bays, or a bridge two hundred and twenty five feet long.

The advance guard equipage is also divided into trains, each train having four ponton divisions. A division contains eight ponton wagons, two wagons for chess, and two for trestles. The ponton wagons of this equipage are so loaded that each wagon will have all the material necessary to construct one complete bay. By this arrangement the number of wagons may be increased or diminished, as the case may require.

220. Bridges on rafts.—A bridge equipage is not always provided, and hence some temporary expedient must be adopted by means of which the stream may be crossed. Or, the equipage may not have materials enough to build as long a bridge as may be required.

Under these circumstances, floating supports may be improvised from casks, or from timbers, if they can be obtained.

When casks are used, they should be laid in line, with the bungs up. Then with two short pieces of scantling to be used as gunwales, the casks may all be firmly fastened together, either by lashing them to the scantlings with ropes, or by framing with timber.

The casks may be arranged in a single row to form the support, or in two rows placed side by side. If in a single row, the lengths of the casks are usually placed at right angles to the direction of the current; if in two rows, the lengths may be in the direction of the current. The floating support thus formed should extend a sufficient distance beyond the ends of the chess to avoid dangerous oscillations.

Rafts of timber are frequently used as floating supports, and are constructed by throwing the logs into the water at places suitable for building the rafts. The logs assume the position they will naturally take in the water. They are then drawn into the bank, and the under side of the end which is to go up stream is trimmed off to present as little obstruction to the current as possible.

The logs are then brought together, occupying the positions they are to have in the raft, with the butts alternately up and down stream, and are fastened together by stout scantlings or poles, placed at right angles to the logs, and spiked or pinned to them. The arrangements are then made to receive the balks and chess as in other bridges.

The largest and longest timber makes the best rafts. It is to be remembered that, in this country, there are but few varieties of timber which possess, when green, the necessary buoyancy for a raft.

221. Bridges with fixed supports.—In bridges of this class, the balks rest on points of support which do not depend upon their buoyant qualities.

Trestles, piles, and crib-work, are the kinds of supports most ordinarily used.

Trestles.—A trestle is a wooden frame, consisting of a horizontal beam, termed the cap or ridge, supported by two or more legs.

A bridge laid upon trestles is easily constructed in shallow-water, and can ordinarily be built of the materials procured in the vicinity. A bridge of this kind is in every way inferior to the ponton bridge, but when this latter is not to be had in time, and the materials for making the trestles are near at hand, it will answer the purpose of crossing a shallow stream, and can be quickly built. The methods used to construct the trestles, to place them in position, and to build the bridge, are given in the manuals for pontoniers.

Piles.—Pile bridges are frequently used in military operations. They are mostly used in the present day, after an army has crossed a stream, to keep open the communications in its rear, and to allow, at the same time, the pontons to be taken up and used at other points.

Crib-work.—In rapid streams with hard bottoms, cribs of rough timber, sunk in the stream and filled with stones, may be used, instead of piles, as the supports of the road-way.

222. Selection of crossings.—A retiring army destroys the bridges over which it passes, to retard the progress of the pursuing forces.

If no opposition is made, the points where bridges previously existed will usually be selected as points of crossing. If opposition is made, it will not be expedient to use these points, but to select others where the bridge can be built without undue exposure of the men, and where it will be practicable to cross over troops in sufficient force and quick enough to hold the ground on the opposite side.

The officer who has the duty of selecting such points must consider the subject technically and tactically.

Tactically, he will look for a point of crossing which will offer advantages for the defence of the bridge, and for protection to the men while constructing it. The considerations which govern him in a tactical sense may be briefly stated as follows: (Art. 163.)

1. The selection of a point which admits of an

approach to it hidden from the enemy's view, and which is also convenient to be used;

2. The selection of a point where the bridge and the approach to it are sheltered from the fire of the enemy;

3. The selection of a point where the banks are higher than, or at least as high as those on the opposite side, and which will afford good positions for artillery;

4. The selection of a point where the ground on the opposite side admits of deployment of troops when they are crossed over, admits of a good defence, and admits of being swept by the artillery fire from the positions just mentioned.

The technical considerations affect the construction of the bridge. These may be briefly stated as follows:

1. The height, the steepness, and the kind, of banks are to be considered;

2. The kind of bottom and the width of the stream are to be considered; and,

3. The strength of the current, and the liability of the stream to freshets are to be considered.

If tactical considerations are the more important, a point in the bend of the stream, is to be preferred. If not, a point on a straight portion of the stream affords the most advantages.

The selection of the point is, in many cases, governed by other circumstances. As, for instance, in the case of insufficient quantity of bridge material, the point may have to be taken where the stream is most narrow.

223. Preservation of bridges.—An officer with a number of men is left in charge of a military bridge to watch it and take care of it.

This officer should see that a depot is established in a safe and convenient place, in which he can have spare materials stored to be used in repairs to the bridge as soon as they are needed.

The officer in charge should inspect the bridge frequently, and examine closely all its parts. Particular attention should be paid to the condition of the cables and to the position of the anchors. In the ponton bridge, he must see that the boats are kept free from water, which might enter through leaks, or be tossed in by waves.

He should keep a guard posted near the bridge, and should place sentinels at each end, or at other points, if necessary. The guard must be instructed as to who are to pass, and who are not; are to see that cavalry dismount and lead their horses; that infantry march at the route step; that beef cattle be allowed to pass over only in small bodies of five or six at a time; that wagons too heavily loaded be kept off the bridge; that in case of the oscillations becoming dangerous, the troops be halted and the artillery stopped until the oscillations cease; and in case of a bridge being used for crossing in both directions, that the right of way be strictly observed.

When a bridge is liable to injury from floating bodies such as may be sent down by the enemy, or may be brought down by freshets, or by other causes, it is necessary to place a guard some distance above the bridge, and to provide them with means to arrest or destroy these hurtful objects. It is recommended to place this guard a half mile or so above the bridge, and furnish them with boats, grapnels, anchors, etc. As soon as the floating objects appear, they may be stopped and towed into secure places; or be so directed as to pass through the bridge without harm to it. In the latter case, a part of the bridge, if it be a ponton bridge, can be made into a **draw**, and this portion removed as the objects approach, thus allowing the floating objects to pass.

The manuals prescribe other methods for protecting a bridge against such dangers. Other means will suggest themselves to the officer in charge.

224. Remark.—There has been no intention in the foregoing paragraphs to give such detailed information in bridge construction as will enable an officer ignorant of bridge-building and of carpentry to build a bridge. It has been intended only to draw attention to an important part of the military art, upon which the success of campaigns frequently depend. Thus, having his attention drawn to the subject, the military student may be led to read and study the works on "Military Bridges."

In organized armies, there are engineers whose business it is to understand bridge-building, and who will have charge of such work. Nevertheless, there are times, even when engineers do have charge of this branch, that officers are placed in positions where streams are to be crossed, and when there are neither engineers, nor bridge equipage at hand. It is under these circumstances, that an officer shows his value and his fitness for his position. The ability he displays in making use of the materials at hand, and the success which follows, will stamp him either as an officer of talent and resources, or as one of little use in an emergency. It has been said that a "staff officer possessed of resource, with the energy necessary to use it properly, may be of more value to an army than the addition of an army corps."

225. Examples.—History records many examples of the use of military bridges. Bridges prepared in advance and carried along; improvised bridges, made from materials found near the place of crossing; bridges resting on fixed, and bridges on floating supports; bridges made hurriedly, and those in which time was not considered, are all more or less mentioned. Disasters arising from want of bridge material, and failures resulting from the same cause, fill pages of military history.

The first military bridge of magnitude, of which we have detailed accounts, was the one built of boats, over the Hellespont, by Xerxes, when he invaded Greece, nearly two thousand four hundred years ago. This bridge was about one and one half (1½) miles long, and was composed of two roadways. One was used by the troops; the other by the baggage train and camp followers. It is said that the number with him was 5,283,220, and that they were seven days and nights in crossing.

Bridges across the Tigris resting on boats are mentioned by Xenophon.

Alexander the Great used skins of animals inflated, or filled with hay, as floats in crossing streams, as shown in his passage of the Oxus.

The Romans carried with their armies small boats and bridge material, when rivers intersected their lines of march.

The first example of a military bridge resting upon fixed points of support, of which we have a detailed account, is the one described in the fourth book of "Cæsar's Commentaries." This bridge was across the Rhine, and was of sufficient strength to meet all the demands made upon it.

History teems with descriptions of military operations along the Rhine, and of the means used to pass this river, by armies operating along its banks. The same may be said with reference to the Danube.

History records, in many cases, the failures arising from a want of a military equipage, and the disasters averted by the use of such equipage, or by the construction of an improvised bridge from the materials found in the neighborhood.

The want of a bridge equipage was particularly felt by Bonaparte in his campaign in Italy, in 1796. The presence of such an equipment would have enabled him to cross the Po in time to place his forces in the rear of the Austrians, and would have avoided the forcing of a passage over the Adda, at Lodi.

The importance of bridge equipages was particularly felt by the armies of the United States in the war of 1861–5. The delay in the arrival of the bridge material was, on more than one occasion, the cause of disaster.

No better example of the importance of a bridge equipage and the value of skilled pontoniers can be given than the single instance of Napoleon's crossing of the Beresina, in 1812, in his retreat from Moscow.

Napoleon proposed to cross the Beresina at Borisov, by a bridge at that point, which was held by his troops. On the 23d of November, he learned that the Russians were in possession of this bridge. In preparing for his retreat, Napoleon had ordered large numbers of wagons and much of the camp equipage to be destroyed, including his bridge equipage.

General Eblé remonstrated, but to no effect. The general, however, made his pontoniers carry with them some of the implements and tools which, he foresaw, would be required in the hasty construction of means for crossing streams over which the bridges would probably be destroyed.

Napoleon sent for General Eblé on the 24th of November, and explained to him his plans. At the point selected for crossing, the river was about one hundred yards wide, six or seven feet deep, with a moderate current, and a muddy bottom.

Had the bridge equipage been saved, the crossing would have been a simple thing, but as it was, a bridge had to be improvised. It was then that the foresight of General Eblé was fully appreciated, for without the tools and few implements he had saved, even the improvised bridge could not have been built.

He decided to construct two trestle bridges; one for the infantry, the other for the artillery and the train. The timber and materials were obtained from trees in the vicinity, and from the demolition of houses.

By 5 P.M. on the 25th of November, he had prepared forty-six trestles. By 1 P.M. on the next day, the bridge for the infantry was finished, and the infantry commenced crossing. By 4 P.M. the other bridge was finished, and the wagons began crossing.

The weather was cold, and there was much floating ice. The sufferings of the pontoniers and their exposure were great. All agree that the pontoniers on this occasion "saved the army."

CHAPTER XVIII.

RAILROADS.

226. Railroads.—Railroads have played an important part in recent wars. Beginning with the Crimean war of 1855, and ending with the late wars in Europe, the military student will be struck with the importance of this class of communications in the efficient supplying of an army, and in the concentration of troops. By their use, numbers are concentrated and supplied in a space of time which was not dreamt of in the beginning of the present century. It is safe to predict that, in all future wars in civilized countries, the railroad will be the line of communication for an army. If a system of railroads already exists, this system will be used; if not, temporary lines of railroad will be constructed. It has now become an important part of an officer's education to understand the principles of construction, and the working, of railroads, to know how they can be preserved, and how they may be destroyed.

227. Construction.—The construction of a railroad for military purposes differs from that intended for peaceful traffic only in the degree of excellence. Economy and rapidity are the essential qualities looked for in the construction of a military railroad. The principal things in its construction are the grading, and the laying of the track.

Grades and curves are necessary evils incident to railroads, and a proper selection of them requires an exercise of good judgment, in many cases. Sometimes, the track may be laid on the natural surface of the ground, or with so little filling and excavation as to amount to nearly the same thing. The placing of the cross-ties, the spiking of the rails, and the general finishing of the road are better done, when men used to this kind of labor can be procured. Usually there can be found among the troops, a great many who have a practical knowledge of this class of construction, and these men can be profitably used as foremen and superintendents of the working parties.

228. Working of railroads.—The successful working of a railroad requires an efficient superintendent, as much as it requires sufficient rolling-stock and good locomotives. A good man for superintendent can generally be obtained from some of the railroad companies, but he has the defect, as a rule, of knowing nothing of the peculiarities of military service. Nevertheless, his experience and knowledge will be of great service to the military officer in charge of the road, and the working may thus be made successful.

From the numbers employed upon the railroads in the United States, there will be no difficulty, in future wars, in the government obtaining as many men as may be necessary, who will be thoroughly cognizant of the duties that may be required of them. In the beginning, there will be some friction and irregularities, but these will wear off, and an efficient corps of trained men can soon be formed. It would be better, however, if "time was taken by the forelock," and a skeleton organization formed in advance.

Engineer officers should pay particular attention to this part of their profession, and be ready, on short notice, to organize bodies of workmen whose special duties will be those assigned to construction, working, and preservation of railroads. And since the other officers of the army are more or less liable to be assigned to duties connected with the preservation, as well as the construction, of these roads, it is equally incumbent on them to acquire this knowledge and be able to put it to a practical use.

229. Movements by railroads.—This subject is really included in the "working of railroads." It is alluded to particularly on account of its importance, especially in moving troops, and is mentioned here simply to advise forethought in those connected with such movements.

The movement of troops by railroad may be divided into five distinct parts, viz.:

1. The march to the point where the troops are to get into the cars.

2. The embarkation.

3. The journey.

4. Leaving the train at the end of the journey.

5. The march from this point to the place of camping.

A careful examination made beforehand of each portion of the movement will add greatly to the soldier's comfort, and prevent much confusion, delay, and annoyance.

Elaborate rules are laid down, both in the Prussian and French services, for moving troops by rail, and it is recommended that these be read by officers who desire to inform themselves on this subject.

230. Preservation of railroads.—The protection of a railroad and guarding it from injury fall within the province of the duties assigned to all officers of the service. Whereas the construction and working of railroads belong to engineers, or a construction corps, the movement of troops to the quartermaster's department, the protection of the road becomes the duty of all branches of service.

A line of railroad used as a line of communication of an army with its base, is protected in a great measure by the army itself. It is, however, liable to injury from cavalry raids of the enemy, and from the acts of a hostile population, if they be present.

The parts most liable to be destroyed or injured are the bridges and tunnels. Guards should be stationed near these points and be protected by field-works or block-houses.

The general track of the road should be carefully watched by trackmen and patrols. Cavalry detachments should scour the approaches in every direction, to give timely notice of approaching raids, and to arrest suspicious persons in the vicinity of the railroad.

The degree of caution and watchfulness needful will largely depend upon the activity of the enemy, who may attempt an interruption of the road by cavalry raids, and will depend also upon the character of the inhabitants living near the road. The latter, when very hostile, may resort to many devices to interrupt the passage of trains, and to injure the bridges.

231. Destruction of railroads.—The destruction of a railroad, or an injury inflicted upon it so that it cannot be quickly repaired, may form, at times, the special duty of any officer.

There are two general cases; one, where the injuries inflicted are to prevent its use by an enemy; and the other, where it is desired to do as much injury as possible, and render the work irreparable, compelling an actual reconstruction of the road.

The first consists in removing parts of the rolling-stock and hiding them, or, where rails cannot be obtained, in removing the track for a few hundred yards or so, at various intervals.

The following is a method of removing the track to render the road temporarily useless :

Select a part of the track laid on a high embankment. Tear up the rails at the extremities of the part to be removed. Line the outside of the track with men for the whole length of the portion to be taken up, and have the men to face inwards. At a given signal the men seize the rail next to them ; and, at another signal, all lift the rail, raising it and the ties to a vertical position, when they let the whole fall over the embankment. If the road is ballasted, the men must provide themselves with levers to lift the track. The portion thrown over the embankment cannot be replaced until the rails are unfastened from the ties, and this takes time.

The second case consists in removing the rails and bending and twisting them so as to render them unfit for use in repairing the road ; in burning or blowing up the bridges ; destroying the tunnels ; disabling the rolling stock, etc.

Locomotives can be temporarily disabled by removing parts of the machinery. They may be permanently disabled by firing a round shot through the boiler. Another way, is to let out all the water in the boiler and then build a large fire in the fire-box ; the fire soon destroys the flues.

All other kinds of rolling-stock may be temporarily disabled by removing parts, or permanently injured by burning them.

Some labor is required to bend and twist the rails, as it is not an easy matter to remove the rails from the ties.

Workmen have special tools for drawing out the spikes and unscrewing the nuts, but these tools are too heavy to be carried upon a raid, where time is so important an element. But when the rails have been taken up, and there is time. it is recommended to form the ties into heaps, and set them on fire. Then to place the rails on the burning heap, loading the ends with other ties. As the rails become red hot, they will bend under the load, and cannot be used again until they are straightened. This bending may also be done by men catching the ends of the rail and bending it, while heated, against a tree or telegraph pole.

Rails which are simply bent can easily be straightened by re-heating and hammering. Where only slightly bent, they can be straightened without even being re-heated.

To make them useless, it is necessary to give the rail a twist. A twisted rail can only be used again by being re-rolled.

This twisting may be effected by a contrivance devised by General Haupt while in charge of the Bureau of Military Bridges, in the War Department, in 1861–5.

The contrivance is a U-shaped piece of tough iron or steel (Fig. 97) which weighs about 6½ lbs., and having its ends turned up into claws.

The method of using it was to take two of these pieces, near together, and force them under the flange near the end of the rail to be twisted and torn off, until the claws caught the lower edge, as shown in

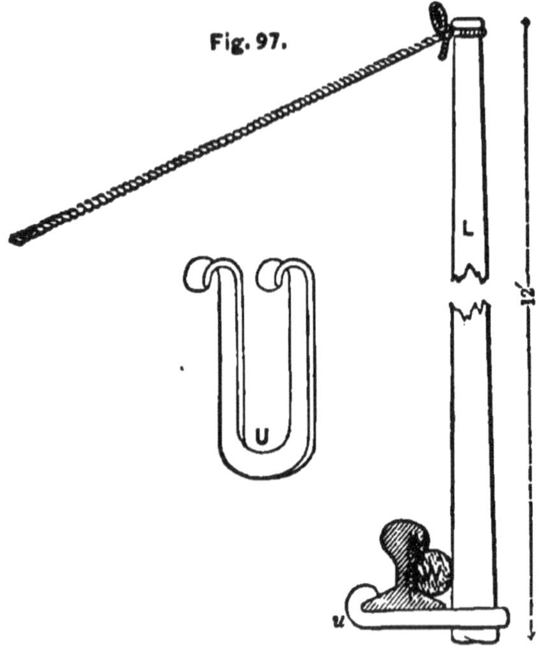

the figure. There was sufficient room between the rail and the loop of the U to insert a lever, L. These levers were obtained in the vicinity, were about twelve feet long, with a cross-section from five to six inches. When inserted in the loop they were firmly held in place by a wedge, W.

The first lever was pulled in and bent to the ground, and the second pulled as far as it could be moved; a second hold was then taken and the lever pulled down as before; this operation was continued until sufficient twist had been given. By this operation, the end of the rail was detached from the ties. A rope was fastened to it, and with the levers, a bend was made in the rail.

A squad of ten men could twist, remove, and bend a rail twenty feet long in five minutes. This would require one hour for a squad to tear up forty yards of track. By increasing the number of squads, it is easy to estimate the running feet of track that could be destroyed in a given time.

Wooden bridges may be destroyed by burning. A simple device called a torpedo was used in our late war for destroying wooden bridges, where time was of importance. A bolt of $\frac{7}{8}$ inch iron, 8 inches long, with head and nut, was used. The head was 2 inches in diameter, and 1 inch thick. A tin cylinder, $1\frac{3}{4}$ inches in diameter, open at both ends, enclosed the bolt and was held in place by the head and the nut. A washer between the head and the cylinder made it tight at that end. The cylinder was filled with powder, and an arrangement made for a fuze near the nut. A fuze was inserted and the nut screwed on, and the torpedo was ready for use.

In using it, a hole was bored into the timber with an auger. The head of the bolt was inserted and driven by a blow into the hole. The fuze was lighted, and the explosion tore the timber in pieces.

As the railroad bridges to be destroyed were ordinary truss-bridges, it was only necessary to insert a torpedo in one of the main braces, or if these braces were in pairs, in the two pieces forming a pair. The destruction of these braces at one end, or on one side, was sufficient to wreck the bridge.

232. Remark.—The importance of guarding a railroad and of having a good construction corps organized to repair the damages, was illustrated in the war of 1861–5. This war illustrated the uses of the systems of railroads already in existence for military purposes, and also the advantages of temporary railroads to perform a given service.

No special allusion is here made to telegraphy. A telegraph line forms an essential element of "working a railroad," and cannot be surpassed as a means of communication for messages and signals.

In closing the subject, it is not to be understood that the details here given are more important than some that are left out. The reverse is the case, in some instances. The "arrangements for the comfort and health of troops" are second to none, and the officer in charge of men, as soon as their immediate safety can be secured from an attack by an enemy, should devote himself to the arrangements which are essential to the health of his command, and which add to their comfort.

END.

www.ingramcontent.com/pod-product-compliance
Lightning Source LLC
Chambersburg PA
CBHW021505210326
41599CB00012B/1141

* 9 7 8 3 7 4 3 3 4 7 7 4 8 *